高水平地方应用型大学建设系列教材

Environmental Chemistry 学习指导教程

蒋路漫　王罗春　周　振　编著

北　京
冶金工业出版社
2022

内 容 提 要

本书采用英中文对照形式,帮助学生在理解环境化学专业知识的基础上,读懂专业英语。全书共分为 5 章,内容包括环境化学与环境五大圈层、大气圈环境化学、水圈环境化学、地圈环境化学、生物圈环境化学。各章内容由知识框架、重要名词和术语解析、本章习题及详解 3 部分组成。

本书可作为高等院校环境科学、环境工程专业英语课教材,也可供相关科研人员参考。

图书在版编目(CIP)数据

Environmental Chemistry 学习指导教程/蒋路漫,王罗春,周振编著. —北京:冶金工业出版社,2022.9
高水平地方应用型大学建设系列教材
ISBN 978-7-5024-9077-5

Ⅰ. ①E… Ⅱ. ①蒋… ②王… ③周… Ⅲ. ①环境化学—英语—高等学校—教材 Ⅳ. ①X13

中国版本图书馆 CIP 数据核字(2022)第 040193 号

Environmental Chemistry 学习指导教程

出版发行	冶金工业出版社	电　话	(010)64027926
地　址	北京市东城区嵩祝院北巷 39 号	邮　编	100009
网　址	www.mip1953.com	电子信箱	service@mip1953.com

责任编辑　刘林烨　程志宏　美术编辑　彭子赫　版式设计　郑小利
责任校对　梅雨晴　责任印制　李玉山

三河市双峰印刷装订有限公司印刷
2022 年 9 月第 1 版,2022 年 9 月第 1 次印刷
710mm×1000mm 1/16;19 印张;368 千字;281 页
定价 40.00 元

投稿电话　(010)64027932　投稿信箱　tougao@cnmip.com.cn
营销中心电话　(010)64044283
冶金工业出版社天猫旗舰店　yjgycbs.tmall.com
(本书如有印装质量问题,本社营销中心负责退换)

《高水平地方应用型大学建设系列教材》
编 委 会

主　任　　徐群杰

副主任　　王罗春　李巧霞

委　员　（按姓氏笔画）

　　　　　　王罗春　王　莹　王　啸　刘永生　任　平
　　　　　　朱　晟　李巧霞　陈东生　辛志玲　吴春华
　　　　　　张俊喜　张　萍　沈喜训　时鹏辉　赵玉增
　　　　　　郑红艾　周　振　孟新静　胡晨燕　高立新
　　　　　　郭文瑶　郭新茹　徐群杰　葛红花　蒋路漫
　　　　　　赖春艳　蔡毅飞

《高水平地方应用型大学建设系列教材》序

应用型大学教育是高等教育结构中的重要组成部分。高水平地方应用型高校在培养复合型人才、服务地方经济发展以及为现代产业体系提供高素质应用型人才方面越来越显现出不可替代的作用。2019年，上海电力大学获批上海市首个高水平地方应用型高校建设试点单位，为学校以能源电力为特色，着力发展清洁安全发电、智能电网和智慧能源管理三大学科，打造专业品牌，增强科研层级，提升专业水平和服务能力提出了更高的要求和发展的动力。清洁安全发电学科汇聚化学工程与工艺、材料科学与工程、材料化学、环境工程、应用化学、新能源科学与工程、能源与动力工程等专业，力求培养出具有创新意识、创新性思维和创新能力的高水平应用型建设者，为煤清洁燃烧和高效利用、水质安全与控制、环境保护、设备安全、新能源开发、储能系统、分布式能源系统等产业，输出合格应用型优秀人才，支撑国家和地方先进电力事业的发展。

教材建设是搞好应用型特色高校建设非常重要的方面。以往应用型大学的本科教学主要使用普通高等教育教学用书，实践证明并不适应在应用型高校教学使用。由于密切结合行业特色及新的生产工艺以及与先进教学实验设备相适应且实践性强的教材稀缺，迫切需要教材改革和创新。编写应用性和实践性强及有行业特色教材，是提高应用型人才培养质量的重要保障。国外一些教育发达国家的基础课教材涉及内容广、应用性强，确实值得我国应用型高校教材编写出版借鉴和参考。

为此，上海电力大学和冶金工业出版社合作共同组织了高水平地方应用型大学建设系列教材的编写，包括课程设计、实践与实习指导、实验指导等各类型的教学用书，首批出版教材 18 种。教材的编写将遵循应用型高校教学特色、学以致用、实践教学的原则，既保证教学内容的完整性、基础性，又强调其应用性，突出产教融合，将教学和学生专业知识和素质能力提升相结合。

本系列教材的出版发行，对于我校高水平地方应用型大学的建设、高素质应用型人才培养具有十分重要的现实意义，也将为教育综合改革提供示范素材。

上海电力大学校长　李和兴

2020 年 4 月

前　言

环境化学是环境学专业重要的专业基础课程之一，为了帮助在校的大学生学好环境化学，编者在参考了 Stanley E. Manahan 教授编写的 *Environmental Chemistry（Tenth edition）*、戴树桂教授主编的《环境化学（第 2 版）》等书籍和有关资料的基础上，融知识体系架构梳理、术语解析和习题讲解为一体编写了本书。

本书每章内容由知识框架、重要名词和术语解析、本章习题及详解 3 部分组成。知识框架部分列出了每章应掌握的基本概念的中英文对照，突出了必须掌握的知识重点；重要名词和术语解析部分对 *Environmental Chemistry（Tenth edition）* 书中提到的重要名词和术语进行列表整理，并给出了中英文解释。为了方便读者自我检测，读者可通过扫描每章习题后的二维码，查阅中英双语详细解答，以帮助读者理清解题思路，掌握基本解题方法和技巧，旨在提高读者的解题能力、分析能力。

本书由蒋路漫、王罗春、周振编写。由于编者水平所限，书中不妥之处，敬请广大读者批评指正。

编　者
2022 年 3 月

目 录

1 Environmental Chemistry and the Five Spheres of the Environment
环境化学与环境五大圈层 ………………………………………………………… 1

1.1 the Earth and the Earth System
地球与地球系统 ………………………………………………………… 1

1.2 Biogeochemical Cycles in the Earth System
地球系统中的生物地球化学循环 ……………………………………… 3

1.3 Natural Capital of the Earth System
地球系统的自然资本 …………………………………………………… 5

1.4 What is Environmental Chemistry?
什么是环境化学？ ……………………………………………………… 6

1.5 Environmental Chemistry of Air and the Atmosphere
大气与大气环境化学 …………………………………………………… 6

1.6 Environmental Chemistry of Water and the Hydrosphere
水与水圈的环境化学 …………………………………………………… 7

1.7 Environmental Chemistry of the Geosphere
地圈的环境化学 ………………………………………………………… 8

1.8 Environmental Chemistry of the Anthrosphere
人类活动圈环境化学 …………………………………………………… 9

1.9 Environmental Chemistry of the Biosphere
生物圈环境化学 ………………………………………………………… 11

Exercises
习题 …………………………………………………………………………… 13

2 Environmental Chemistry of the Atmosphere
大气圈环境化学 ……………………………………………………………… 16

2.1 The Atmosphere and Atmospheric Chemistry
大气与大气化学 ………………………………………………………… 16

2.1.1 The Atmospheric Composition and the Main Pollutants
大气组成与主要污染物 ·· 16

2.1.2 How the Atmosphere Got That Way and its Natural Capital
大气如何发展及其自然资本 ·· 18

2.1.3 Physical Characteristics of the Atmosphere
大气的物理特征 ·· 19

2.1.4 Energy Transfer in the Atmosphere
大气中的能量转移 ··· 20

2.1.5 Atmospheric Mass Transfer, Meteorology and Weather
大气传质、气象和天气 ··· 21

2.1.6 Inversions and Air Pollution
逆温与空气污染 ·· 25

2.1.7 Global Climate and Microclimate
全球气候与微气候 ··· 25

2.1.8 Chemical and Photochemical Reactions
化学和光化学反应 ··· 27

2.1.9 Acid-Base Reactions in the Atmosphere
大气中的酸碱反应 ··· 31

2.1.10 Reactions of Atmospheric Oxygen
大气中的氧的反应 ··· 32

2.1.11 Reactions of Atmospheric Nitrogen
大气中的氮的反应 ··· 32

2.1.12 Atmospheric Water
大气中的水 ·· 33

2.1.13 Influence of the Anthrosphere
人类活动圈的影响 ··· 33

2.1.14 Chemical Fate and Transport in the Atmosphere
大气中的化学归宿和迁移 ··· 34

2.2 Particles in the Atmosphere
大气颗粒物 ·· 34

2.2.1 Particles in the Atmosphere
大气中的粒子 ··· 34

2.2.2 Physical Behavior of Atmospheric Particles
大气颗粒的物理行为 ·· 36

2.2.3　Physical Processes for Particle Formation
粒子形成的物理过程 …………………………………………… 37

2.2.4　Chemical Processes for Particle Formation
颗粒形成的化学过程 …………………………………………… 38

2.2.5　The Composition of Inorganic Particles
无机颗粒的组成 ………………………………………………… 39

2.2.6　Toxic Metals in the Atmosphere
大气中的有毒金属 ……………………………………………… 40

2.2.7　Radioactive Particles
放射性粒子 ……………………………………………………… 41

2.2.8　Organic Particles in the Atmosphere
大气中的有机粒子 ……………………………………………… 41

2.2.9　Effects of Particles
粒子的影响 ……………………………………………………… 42

2.2.10　Water as Particulate Matter
水作为颗粒物 …………………………………………………… 43

2.2.11　Control of Particle Emissions
颗粒物排放的控制 ……………………………………………… 43

2.3　Gaseous Inorganic Pollutants
气态无机污染物 ………………………………………………………… 45

2.3.1　Inorganic Pollutant Gases
无机污染物气体 ………………………………………………… 45

2.3.2　Production and Control of Carbon Monoxide
一氧化碳的产生与控制 ………………………………………… 45

2.3.3　Fate of Atmospheric CO
大气中一氧化碳的归宿 ………………………………………… 46

2.3.4　Sulfur dioxide Sources and the Sulfur Cycle
二氧化硫的来源和硫循环 ……………………………………… 46

2.3.5　Sulfur Dioxide Reactions in the Atmosphere
大气中的二氧化硫反应 ………………………………………… 47

2.3.6　Nitrogen Oxides in the Atmosphere
大气中的氮氧化物 ……………………………………………… 49

2.3.7　Acid Rain
酸雨 ……………………………………………………………… 51

- 2.3.8 Ammonia in the Atmosphere
  大气中的氨 ···················· 51
- 2.3.9 Fluorine, Chlorine and their Gaseous Compounds
  氟、氯及其气态化合物 ············ 52
- 2.3.10 Reduced Sulfur Gases
  还原的硫气体 ·················· 53
- 2.4 Organic Air Pollutants
  有机大气污染物 ················ 53
  - 2.4.1 Organic Compounds in the Atmosphere
    大气中的有机化合物 ············ 53
  - 2.4.2 Biogenic Organic Compounds in the Atmosphere
    大气中的生物有机化合物 ········ 54
  - 2.4.3 Pollutant Hydrocarbons
    碳氢化合物污染物 ············ 55
- 2.5 Photochemical Smog
  光化学烟雾 ···················· 59
  - 2.5.1 Introduction
    简介 ···················· 59
  - 2.5.2 Smog-Forming Emissions
    烟雾形成排放 ················ 59
  - 2.5.3 Smog-Forming Reactions of Organic Compounds in the Atmosphere
    大气中有机化合物的烟雾形成反应 ···· 62
  - 2.5.4 Overview of Photochemical Smog Formation
    光化学烟雾形成概述 ············ 64
  - 2.5.5 Mechanisms of Smog Formation
    烟雾形成机理 ················ 64
  - 2.5.6 Reactivity of Hydrocarbons
    碳氢化合物的反应性 ············ 68
  - 2.5.7 Importance of HO_x/VOC Ratios
    HO_x/VOC 比的重要性 ·········· 69
  - 2.5.8 Inorganic Products from Smog
    烟雾的无机产物 ··············· 69
  - 2.5.9 Effects of Smog
    烟雾的影响 ·················· 70

2.6 The Endangered Global Atmosphere
　　面临危险的全球大气 ·· 70

　2.6.1 Saving the Atmosphere to Save Ourselves
　　　　拯救大气，拯救我们自己 ······································ 70

　2.6.2 the Earth's Evolving Atmosphere and Climate Change
　　　　地球不断演变的大气与气候变化 ································ 71

　2.6.3 Effects of the Anthrosphere on Atmosphere and Climate
　　　　人类活动圈对大气和气候的影响 ································ 72

　2.6.4 The Greatest Threat to the Atmosphere and the Earth System:
　　　　Global Warming
　　　　对大气和地球系统的最大威胁：全球变暖 ······················· 72

　2.6.5 Green Science and Technology to Alleviate Global Warming
　　　　缓解全球变暖的绿色科学技术 ···································· 75

　2.6.6 Acid Rain
　　　　酸雨 ·· 76

　2.6.7 Stratospheric Ozone Destruction
　　　　平流层臭氧破坏 ··· 78

　2.6.8 Atmospheric Brown Clouds
　　　　大气棕云 ··· 81

　2.6.9 Atmospheric Damage by Photochemical Smog
　　　　光化学烟雾对大气的损害 ·· 82

　2.6.10 The Urban Aerosol
　　　　　城市气溶胶 ·· 84

　2.6.11 What can be done?
　　　　　我们能做什么？ ··· 85

Exercises
习题 ·· 87

3 Environmental Chemistry of the Hydrosphere
水圈环境化学 ··· 109

3.1 The Hydrosphere and Water Chemistry
　　水圈和水化学 ·· 109

　3.1.1 Water: An Essential Part of Earth's Natural Capital
　　　　水：地球自然资本的重要组成部分 ···························· 109

3.1.2　Sources and Uses of Water
水的来源和用途 ………………………………………………… 110

3.1.3　H_2O: Simple Formula, Remarkable Molecule
H_2O: 简单的化学式，杰出的分子 ……………………………… 111

3.1.4　Life in Water (Biota)
水中生物（生物群）……………………………………………… 113

3.1.5　Introduction to Aquatic Chemistry
水化学简介 ……………………………………………………… 115

3.1.6　Gases in Water
水中的气体 ……………………………………………………… 115

3.1.7　Water Acidity and Carbon Dioxide in Water
水中的酸度和二氧化碳 ………………………………………… 116

3.1.8　Alkalinity
碱度 ……………………………………………………………… 118

3.1.9　Calcium and Other Metal Ions in Water
水中的钙和其他金属离子 ……………………………………… 119

3.1.10　Complexation and Chelation
络合与螯合 ……………………………………………………… 120

3.1.11　Bonding and Structure of Metal Complexes
金属配合物的键合与结构 ……………………………………… 122

3.1.12　Calculations of Species Concentrations
物种浓度的计算 ………………………………………………… 122

3.1.13　Complexation by Deprotonated Ligands
去质子配体的络合 ……………………………………………… 123

3.1.14　Complexation by Protonated Ligands
质子化配体的络合 ……………………………………………… 123

3.1.15　Solubilization of Lead Ion from Solids by NTA
NTA 从固体中溶解铅离子 ……………………………………… 124

3.1.16　Polyphosphates and Phosphonates in Water
水中的多磷酸盐和磷酸盐 ……………………………………… 126

3.1.17　Complexation by Humic Substances
腐殖质的络合 …………………………………………………… 127

3.1.18　Complexation and Redox Processes
络合和氧化还原过程 …………………………………………… 128

3.2　Oxidation/Reduction in Aquatic Chemistry
　　水化学中的氧化/还原 ………………………………………………………… 129

　3.2.1　The Significance of Oxidation/Reduction
　　　　氧化/还原的意义 …………………………………………………… 129

　3.2.2　The Electron and Redox Reactions
　　　　电子和氧化还原反应 ………………………………………………… 130

　3.2.3　Electron Activity and pE
　　　　电子活度和 pE ……………………………………………………… 131

　3.2.4　The Nernst Equation
　　　　能斯特方程 …………………………………………………………… 132

　3.2.5　Reaction Tendency: Whole Reaction from Half-Reactions
　　　　反应倾向：半反应引起的整个反应 ………………………………… 132

　3.2.6　The Nernst Equation and Chemical Equilibrium
　　　　能斯特方程和化学平衡 ……………………………………………… 133

　3.2.7　The Relationship of pE to Free Energy
　　　　pE 与自由能的关系 ………………………………………………… 134

　3.2.8　Reactions in Terms of One Electron-Mole
　　　　一个电子摩尔的反应 ………………………………………………… 134

　3.2.9　The Limits of pE in Water
　　　　水中 pE 的限值 ……………………………………………………… 135

　3.2.10　pE Value in Natural Water Systems
　　　　　天然水体系中的 pE 值 …………………………………………… 135

　3.2.11　pE-pH Diagrams
　　　　　pE-pH 图 …………………………………………………………… 136

　3.2.12　Humic Substances as Natural Reductants
　　　　　腐殖质作为天然还原剂 …………………………………………… 137

　3.2.13　Photochemical Processes in Oxidation/Reduction
　　　　　氧化/还原中的光化学过程 ………………………………………… 137

　3.2.14　Corrosion
　　　　　腐蚀 …………………………………………………………………… 137

3.3　Phase Interactions in Aquatic Chemistry
　　水化学中的相间作用 ………………………………………………………… 138

　3.3.1　Importance and Formation of Sediments
　　　　沉积物的重要性和形成 ……………………………………………… 138

3.3.2　Solubilities
　　　溶解度 …………………………………………………………………… 139

3.3.3　Colloidal Particles in Water
　　　水中的胶体颗粒 ………………………………………………………… 141

3.3.4　The Colloidal Properties of Clays
　　　黏土的胶体性质 ………………………………………………………… 144

3.3.5　Aggregation of Particles
　　　粒子聚集 ………………………………………………………………… 145

3.3.6　Surface Sorption by Solids
　　　固体表面吸附 …………………………………………………………… 146

3.3.7　Solute Exchange with Bottom Sediments
　　　与底部沉积物的溶质交换 ……………………………………………… 147

3.3.8　Interstitial Water
　　　间隙水 …………………………………………………………………… 150

3.3.9　Phase Interactions in Chemical Fate and Transport
　　　化学归宿和迁移中的相间作用 ………………………………………… 151

3.4　Aquatic Microbial Biochemistry
　　水生微生物生物化学 ……………………………………………………… 152

3.4.1　Aquatic Biochemical Processes
　　　水生生化过程 …………………………………………………………… 152

3.4.2　Algae
　　　藻类 ……………………………………………………………………… 154

3.4.3　Fungi
　　　真菌 ……………………………………………………………………… 155

3.4.4　Protozoa
　　　原生动物 ………………………………………………………………… 156

3.4.5　Bacteria
　　　细菌 ……………………………………………………………………… 156

3.4.6　The Prokaryotic Bacterial Cell
　　　原核细菌细胞 …………………………………………………………… 158

3.4.7　Kinetics of Bacterial Growth
　　　细菌生长动力学 ………………………………………………………… 159

3.4.8　Bacterial Metabolism to Get Energy and Cellular Material
　　　通过细菌代谢获取能量和细胞物质 …………………………………… 160

3.4.9 Microbial Transformations of Carbon
碳的微生物转化 ………………………………………………………… 161

3.4.10 Biodegradation of Organic Matter
有机物的生物降解 ……………………………………………………… 162

3.4.11 Microbial Transformations of Nitrogen
氮的微生物转化 ………………………………………………………… 164

3.4.12 Microbial Transformations of Phosphorus and Sulfur
磷和硫的微生物转化 …………………………………………………… 165

3.4.13 Microbial Transformations of Halogens and Organohalides
卤素和有机卤化物的微生物转化 ……………………………………… 166

3.4.14 Microbial Transformations of Metals and Metalloids
金属和准金属的微生物转化 …………………………………………… 167

3.5 Water Pollutants and Water Pollution
水污染物和水污染 …………………………………………………………… 169

3.5.1 Nature and Types of Water Pollutants
水污染物的性质和类型 ………………………………………………… 169

3.5.2 Elemental Pollutants
元素污染物 ……………………………………………………………… 170

3.5.3 Heavy Metals
重金属 …………………………………………………………………… 171

3.5.4 Metalloids
类金属 …………………………………………………………………… 172

3.5.5 Organically Bound Metals and Metalloids
有机结合的金属和准金属 ……………………………………………… 172

3.5.6 Inorganic Species
无机物 …………………………………………………………………… 173

3.5.7 Algal Nutrients and Eutrophication
藻类营养物和富营养化 ………………………………………………… 175

3.5.8 Acidity, Alkalinity and Salinity
酸度、碱度和盐度 ……………………………………………………… 176

3.5.9 Oxygen, Oxidants and Reductants
氧、氧化剂和还原剂 …………………………………………………… 176

3.5.10 Organic Pollutants
有机污染物 ……………………………………………………………… 177

3.5.11 Pesticides in Water
水中的农药 ········· 181

3.5.12 Organochlorine Compounds in Water
水中的有机氯化合物 ········· 186

3.5.13 Emerging Water Pollutants, Pharmaceuticals and Household Wastes
新兴水污染物、药品和生活垃圾 ········· 188

3.5.14 Radionuclides in the Aquatic Environment
水生环境中的放射性核素 ········· 189

Exercises
习题 ········· 191

4 Environmental Chemistry of the Geosphere
地圈环境化学 ········· 211

4.1 Geosphere and Geochemistry
地圈和地球化学 ········· 211

4.1.1 The Fragile Solid Earth and its Relationship with the Other Environmental Spheres
脆弱的固体地球及其与其他环境圈层的关系 ········· 211

4.1.2 The Geosphere as a Source of Natural Capital
地圈作为自然资本的来源 ········· 212

4.1.3 Effects of Human Activities on the Geosphere
人类活动对地圈的影响 ········· 213

4.1.4 Air Pollution and the Geosphere
空气污染与地圈 ········· 213

4.1.5 Water Pollution and the Geosphere
水污染与地圈 ········· 213

4.2 Soil: Earth's Lifeline
土壤：地球的生命线 ········· 214

4.2.1 Have You Thanked a Clod Today?
今天你感谢土块了吗? ········· 214

4.2.2 Structure of Soil
土壤结构 ········· 215

4.2.3 Composition of Soil
土壤成分 ········· 216

4.2.4　Acid-Base and Ion-Exchange Reactions in Soil
　　　土壤中的酸碱和离子交换反应 …………………………………… 217

4.2.5　Macronutrients in Soil
　　　土壤中的大量养分 ………………………………………………… 219

4.2.6　Nitrogen, Phosphorus and Potassium in Soil
　　　土壤中的氮、磷和钾 ……………………………………………… 219

4.2.7　Micronutrients in Soil
　　　土壤中的微量营养素 ……………………………………………… 220

4.2.8　Fertilizers
　　　肥料 ………………………………………………………………… 221

4.2.9　Pesticides and their Residues in Soil
　　　土壤中的农药及其残留 …………………………………………… 224

4.2.10　Wastes and Pollutants in Soil
　　　　土壤中的废物和污染物 ………………………………………… 224

4.2.11　Soil Loss and Degradation
　　　　水土流失和退化 ………………………………………………… 226

4.2.12　Saving the Land
　　　　救治土地 ………………………………………………………… 228

4.2.13　Green Chemistry and Sustainable Agriculture
　　　　绿色化学与可持续农业 ………………………………………… 229

4.2.14　Genetics and Agriculture
　　　　遗传与农业 ……………………………………………………… 230

4.2.15　Agriculture and Health
　　　　农业与健康 ……………………………………………………… 231

4.2.16　Protecting the Food Supply from Attack
　　　　保护粮食供应不受攻击 ………………………………………… 231

Exercises
习题 ………………………………………………………………………………… 232

5　Environmental Chemistry of the Biosphere
生物圈环境化学 ………………………………………………………… 236

5.1　The Biosphere: Environmental Biochemistry
　　生物圈：环境生物化学 ……………………………………………………… 236

5.1.1　Life and the Biosphere
　　　生命和生物圈 ……………………………………………………… 236

5.1.2　Metabolism and Control in Organisms
　　　　生物体内的代谢与控制 ………………………………………… 238
5.1.3　Reproduction and Inherited Traits
　　　　繁殖与遗传性状 ………………………………………………… 240
5.1.4　Stability and Equilibrium of Biosphere
　　　　生物圈的稳定和平衡 …………………………………………… 240
5.1.5　Biochemistry
　　　　生物化学 ………………………………………………………… 241
5.1.6　Biochemistry and the Cell
　　　　生物化学和细胞 ………………………………………………… 243
5.1.7　Proteins
　　　　蛋白质 …………………………………………………………… 246
5.1.8　Carbohydrates
　　　　碳水化合物 ……………………………………………………… 249
5.1.9　Lipids
　　　　脂质 ……………………………………………………………… 250
5.1.10　Enzymes
　　　　酶 ………………………………………………………………… 252
5.1.11　Nucleic Acids
　　　　核酸 ……………………………………………………………… 254
5.1.12　Recombinant DNA and Genetic Engineering
　　　　重组 DNA 与基因工程 ………………………………………… 256
5.1.13　Metabolic Processes
　　　　代谢过程 ………………………………………………………… 257
5.1.14　Metabolism of Xenobiotic Substances
　　　　异种物质的代谢 ………………………………………………… 259
5.2　Toxicological Chemistry
　　毒理化学 …………………………………………………………… 261
5.2.1　Introduction to Toxicology and Toxicological Chemistry
　　　　毒理学与毒理化学概论 ………………………………………… 261
5.2.2　Dose-Response Relationships
　　　　剂量反应关系 …………………………………………………… 264
5.2.3　Relative Toxicities
　　　　相对毒性 ………………………………………………………… 264

- 5.2.4 Reversibility and Sensitivity
  可逆性和敏感性 ………………………………………………… 265
- 5.2.5 Xenobiotic and Endogenous Substances
  异源和内源性物质 ……………………………………………… 266
- 5.2.6 Toxicological Chemistry
  毒理化学 ………………………………………………………… 266
- 5.2.7 Kinetic Phase and Dynamic Phase
  动力学阶段和动态阶段 ………………………………………… 268
- 5.2.8 Teratogenesis, Mutagenesis, Carcinogenesis, and Effects on the Immune and Reproductive Systems
  致畸、致突变、致癌作用及其对免疫和生殖系统的影响 ……… 269
- 5.2.9 Health Hazards
  健康危害 ………………………………………………………… 275

Exercises
习题 …………………………………………………………………… 276

参考文献 ………………………………………………………………… 281

1 Environmental Chemistry and the Five Spheres of the Environment
环境化学与环境五大圈层

1.1 the Earth and the Earth System
地球与地球系统

知识框架

Earth's environment consists of five closely related and interacting spheres

(1) The hydrosphere: Water.
(2) The atmosphere: Air.
(3) The geosphere: Rock, mineral matter, soil.
(4) The anthrosphere: Things humans make and do.
(5) Biosphere: Living organisms.

These five spheres compose the Earth System.

Earth's two great fluids that circulate and exchange matter and energy

(1) Surface water, especially in the oceans and rivers.
(2) Air in the atmosphere.

The two great fluids are important driving forces behind biogeochemical cycles.

地球环境由五个紧密相关且相互作用的圈层组成

(1) 水圈：水。
(2) 大气圈：空气。
(3) 地圈：岩石、矿物质、土壤。
(4) 人类活动圈：人类制造的物品和做的事情。
(5) 生物圈：生物体。

这五个圈层构成了地球系统。

地球的循环和交换物质与能量的两种大的流动

(1) 地表水，特别是在海洋和河流中。
(2) 大气中的空气。

两种大的流动是生物地球化学循环背后的重要驱动力。

重要名词和术语解析

<u>Environmental science</u>: Environmental science

<u>环境科学</u>：广义的环境科学是

in its broadest sense is the science of the complex interactions that occur among the terrestrial, atmospheric, aquatic, living, and anthropological environments. Environmental science will be defined as *the study of the earth*, *air*, *water*, *and living environments*, *and the effects of technology thereon*.

Atmosphere: The **atmosphere** is the thin layer of gases that cover Earth's surface.

Hydrosphere: The **hydrosphere** contains Earth's water.

Geosphere: The **geosphere** consists of the solid earth, including soil, which supports most plant life. The part of the geosphere that is directly involved with environmental processes through contact with the atmosphere, the hydrosphere, and living things is the solid **lithosphere**.

Crust: The most important part of it insofar as interactions with the other spheres of the environment are concerned is its thin outer skin composed largely of lighter silicate-based minerals and called the **crust**.

Biosphere: All living entities on Earth compose the **biosphere**. Living organisms and the aspects of the environment pertaining directly to them are called **biotic**, and other portions of the environment are **abiotic**.

Biogeochemical cycle: To a large extent, the strong interactions among living organisms and the various spheres of the abiotic environment are best described by cycles of matter that involve biological, chemical, and geological processes and phenomena. Such cycles are called **biogeochemical cycles**.

大气层：**大气层**是覆盖地球表面的薄薄一层气体。

水圈：**水圈**包含地球上的水。

地圈：**地圈**由包含支持大部分植物生命的土壤在内的固体土地组成。通过与大气、水圈和生物接触而直接参与环境过程的一部分地圈称为**岩石圈**。

地壳：就与环境其他领域的相互作用而言，最重要的部分是其薄的外皮，主要由较轻的硅酸盐类矿物组成，被称为**地壳**。

生物圈：地球上所有的生物实体都构成了**生物圈**。生物体以及与之直接相关的环境被称为**生物**，而环境的其他部分则为**非生物**。

生物地球化学循环：很大程度上，涉及**生物**、化学和地质过程与现象的物质循环可以最好地描述生物体与非生物环境各个领域之间的强烈相互作用。这样的循环称为**生物地球化学循环**。

Environmental chemistry: Environmental chemistry may be defined as the study of the sources, reactions, transport, effects, and fates of chemical species in water, soil, air, and living environments, and the effects of technology thereon.

Environmental biochemistry: The discipline that deals specifically with the effects of environmental chemical species on life is environmental biochemistry.

Toxicological chemistry: Toxicological chemistry is the chemistry of toxic substances with emphasis upon their interactions with biologic tissue and living organisms.

环境化学：环境化学可以定义为对水、土壤、空气和生存环境中化学物质的来源、反应、迁移、作用和归宿以及技术影响的研究。

环境生物化学：专门研究环境化学物种对生命的影响的学科称为环境生物化学。

毒理化学：毒理化学是有毒物质的化学，重点在于它们与生物组织和生物体的相互作用。

1.2 Biogeochemical Cycles in the Earth System
地球系统中的生物地球化学循环

知识框架

Biogeochemical cycles involve circulation and placement of matter in the Earth System.

(1) Often based upon specific elements such as those that are nutrients for organisms.

(2) Most have an atmospheric component; phosphorus cycle is one that does not.

(3) Many are involved with the anthrosphere, for example, nitrogen from the atmosphere bound as nitrogen compounds by industrial chemical synthesis and released to the nitrogen cycle.

Some Important Biogeochemical Cycles

(1) Oxygen cycle: Vast reservoir in atmosphere, present in water and rocks, O_2 released by photosynthesis, taken up by respiration.

生物地球化学循环涉及物质在地球系统中的循环和放置。

(1) 通常基于特定元素，例如构成生物营养物的元素。

(2) 大多数都具有大气成分（除了磷循环）。

(3) 许多气体都与人类活动圈有关，如大气中的氮通过工业化学合成，以氮的化合物的形式释放到氮循环中。

一些重要的生物地球化学循环

(1) 氧循环：大量储存于大气中，也存在于水和岩石中，光合作用释放出的 O_2 被呼吸作用吸收。

(2) Nitrogen cycle: Vast amount of N_2 gas in the atmosphere, bound in N compounds by microorganisms and industrial processes, essential plant nutrient, released back to the atmosphere by denitrification.

(3) Sulfur cycle: In atmosphere as SO_2 and sulfate aerosols, in many minerals, sulfur compounds acted upon by bacteria, in proteins in the biosphere.

(4) Phosphorous cycle: Endogenous without an atmospheric component, essential in biomolecules such as nucleic acids, plant nutrient crucial for food crops, excess in water can cause eutrophication, a pollution condition in water.

（2）氮循环：大气中大量的 N_2 气体，通过微生物和工业过程结合于氮的化合物，是必需的植物养分，通过反硝化释放回大气中。

（3）硫循环：在大气中以 SO_2 和硫酸盐气溶胶存在，在许多矿物质中，细菌作用于生物圈中蛋白质中的硫化合物。

（4）磷循环：内源性的，没有大气成分，是生物分子（如核酸），对粮食作物至关重要的植物养分所必需的，水中过量会引起富营养化，这是水中的污染状况。

重要名词和术语解析

Cycles of matter: Cycles of matter often based on elemental cycles, are of utmost importance in the environment.

Mineralization: biologically bound elements are returned to inorganic states.

Endogenic cycle and **exogenic cycle**: Materials cycles may be divided broadly between **endogenic cycles**, which predominantly involve subsurface rocks of various kinds, and **exogenic cycles**, which occur largely on Earth's surface and usually have an atmospheric component.

Salt solution: All sedimentary cycles involve **salt solutions** or **soil solutions** that contain dissolved substances leached from weathered minerals; these substances may be deposited as mineral formations, or they may be taken up by organisms as nutrients.

物质循环：物质循环通常基于元素循环，在环境中至关重要。

矿化：生物结合的元素恢复为无机态。

内生循环和外生循环：物质循环可以大致分为**内生循环**和**外生循环**，内生循环主要涉及各种地下岩石，外生循环主要发生在地球表面，通常具有大气成分。

盐溶液：所有沉积循环都涉及**盐溶液**或**土壤溶液**，其中包含从风化矿物中浸出的溶解物质；这些物质可能以矿物质的形式沉积，也可能被有机体吸收为营养物质。

Carbon Cycle: Mineral carbon is held in a reservoir of limestone, $CaCO_3$, from which it may be leached into a mineral solution as dissolved hydrogen carbonate ion, HCO_3^-, formed when dissolved $CO_2(aq)$ reacts with $CaCO_3$. In the atmosphere carbon is present as carbon dioxide, CO_2. Atmospheric carbon dioxide is fixed as organic matter by photosynthesis, and organic carbon is released as CO_2 by microbial decay of organic matter.

碳循环：矿物质碳保存在石灰石 $CaCO_3$ 的储层中，从 $CaCO_3$ 浸出至矿物质溶液中，当溶解的 $CO_2(aq)$ 与 $CaCO_3$ 反应时，形成溶解的碳酸氢根离子 HCO_3^-。在大气中，碳以二氧化碳的形式存在。大气中的二氧化碳通过光合作用固定为有机物，而有机碳则通过有机物的微生物分解以 CO_2 的形式释放。

1.3 Natural Capital of the Earth System
地球系统的自然资本

知识框架

Earth's natural capital is its capacity to provide materials, protection, and conditions conducive to life.

(1) Natural resources, such as minerals, soil, water.

(2) Ecosystem services, which include:

1) Hospitable climate.

2) Waste assimilation.

3) Protection from ultraviolet radiation.

Economic systems must evolve that provide good living standards while they:

(1) Increase well-being.

(2) Increase wealth and capital.

(3) Increase efficiency of production.

地球的自然资本是其提供有益于生命的物质、保护和条件的能力。

(1) 自然资源，如矿物质、土壤、水。

(2) 生态系统服务，包括：

1) 宜居的气候；

2) 废物同化；

3) 防紫外线。

经济系统必须不断发展，以提供良好的生活水平：

(1) 增加幸福感；

(2) 增加财富和资本；

(3) 提高生产效率；

(4) Conserve resources while reducing wastes.

Economic systems must emphasize provision of services rather than material goods that deplete resources.

(1) Emulate nature's systems.

(2) Apply principles of green chemistry and industrial ecology.

（4）在节省资源的同时减少浪费。

经济制度必须强调提供服务，而不是消耗资源的物质商品。

（1）模拟自然系统。

（2）应用绿色化学和工业生态学原理。

1.4　What is Environmental Chemistry? 什么是环境化学？

知识框架

(1) Environmental Chemistry and the Spheres of the Earth System.

(2) Environmental chemistry describes the origin, transport, reactions, effects, and fates of chemical species in the five spheres of the Earth System.

（1）环境化学与地球系统圈层。

（2）环境化学描述了地球系统五个圈层中化学物种的来源、迁移、反应、影响和归宿。

1.5　Environmental Chemistry of Air and the Atmosphere 大气与大气环境化学

知识框架

Air and the Atmosphere

(1) Source of essential gases.

1) Oxygen for animals and other organisms.

2) Carbon dioxide for plant photosynthesis.

空气与大气

（1）必需气体的来源。

1）动物和其他生物的氧气。

2）用于植物光合作用的二氧化碳。

3) Nitrogen converted to chemically combined form as a plant nutrient.

4) Oxygen, nitrogen and argon for industrial uses.

(2) Protective functions.

1) Filters out damaging ultraviolet radiation.

2) Regulates Earth's surface temperature within a range compatible with life.

Aspects of Atmospheric Science

(1) Movement of air masses.

(2) Heat balance.

(3) Chemical properties and reactions.

3) 氮转化为化学结合形式作为植物营养物。

4) 工业用氧气、氮气和氩气。

(2) 保护功能。

1) 滤除有害的紫外线。

2) 在与生命兼容的范围内调节地球的表面温度。

大气科学的各个方面

(1) 空气运动。

(2) 热平衡。

(3) 化学性质和反应。

重要名词和术语解析

Atmospheric science: Atmospheric science deals with the movement of air masses in the atmosphere, atmospheric heat balance, and atmospheric chemical composition and reactions.

大气科学：大气科学涉及空气在大气中的运动、大气热平衡以及大气化学成分和反应。

1.6　Environmental Chemistry of Water and the Hydrosphere
　　　水与水圈的环境化学

知识框架

(1) Water Pollution.

(2) Water Treatment.

Many important chemical and biochemical processes occur in water.

Pollution of water is a concern, for example by biodegradable oxygen-depleting contaminants.

Water treatment processes are especially important with growing practice of water recycling.

(1) 水污染。

(2) 水处理。

水中发生许多重要的化学和生化过程。

水的污染令人担忧，例如可生物降解的耗氧污染物。

随着越来越多的水循环实践，水处理工艺尤为重要。

1.7 Environmental Chemistry of the Geosphere
地圈的环境化学

知识框架

The Geosphere and Natural Capital	地圈和自然资本
Soil	土壤
Chemical Composition of Soil	土壤的化学成分

Earth: The Geosphere

Crust is Earth's thin outer skin (5~40km thick).

(1) Interacts with other spheres.

(2) Provides life support, food, minerals, fuels.

Geology is the science of the geosphere

(1) Considers mineral solids.

(2) Interaction with water.

(3) Interaction with atmosphere.

(4) Effects upon and by living organisms.

(5) Engineering geology considers human interactions with and modifications of the geosphere.

Geochemistry describes chemical phenomena in the geosphere

Geosphere as a source of Earth's natural capital.

(1) Soil and plant nutrients.

(2) Minerals.

(3) Fuels.

地球：地圈

地壳是地球的薄薄的外皮（5~40km 厚）。

(1) 与其他领域互动。

(2) 提供生命支持、食物、矿物质、燃料。

地质学是地圈科学

(1) 考虑矿物固体。

(2) 与水的相互作用。

(3) 与大气的相互作用。

(4) 对生物的影响。

(5) 工程地质考虑了人类与地圈的相互作用和对地圈的修改。

地球化学描述了地圈中的化学现象

地圈是地球自然资本的来源。

(1) 土壤和植物养分。

(2) 矿物质。

(3) 燃料。

重要名词和术语解析

Geosphere: The **geosphere**, or solid Earth, is that part of the Earth upon which humans live and from which they extract most of their food, minerals, and fuels.

Crust: Environmental science is most concerned with the **lithosphere**, which consists of the outer mantle and **crust**. The latter is the earth's outer skin that is accessible to humans.

Geology: Geology is the science of the geosphere.

Soil: The most important part of the geosphere for life on earth is **soil** formed by the disintegrative weathering action of physical, geochemical, and biological processes on rock.

地圈：**地圈**是人类生活于地球的部分，人类从中提取大部分食物、矿物质和燃料。

地壳：环境科学最关注**岩石圈**，它由外地幔和**地壳**组成。地壳是人类可以接触到的地球的外皮。

地质学：地质学是地圈科学。

土壤：地圈中对地球生物最重要的部分是**土壤**，它是由物理、地质化学和生物过程在岩石上的分解性风化作用形成的。

1.8 Environmental Chemistry of the Anthrosphere 人类活动圈环境化学

知识框架

The anthrosphere is a potential source of pollutants that contaminate the Earth System.

人类活动圈是污染地球系统的潜在污染源。

Industrial Ecology in the Anthrosphere

(1) Industrial ecology views an industrial system like an industrial ecosystem.

(2) Practice of efficient industrial metabolism.

(3) Enterprises interact to make use of what would be wastes from other enterprises.

(4) Optimizes utilization of resources, energy, and capital.

人类活动圈中的工业生态

(1) 工业生态学将工业系统视为工业生态系统。

(2) 有效的工业代谢实践。

(3) 企业进行交互以利用来自其他企业的废物。

(4) 优化资源、能源和资本的利用。

Green Chemistry in the Anthrosphere

(1) Practice of chemical science and manufacturing within a framework of industrial ecology.

(2) Sustainable, safe, and non-polluting.

(3) Consumes minimum amounts of materials and energy.

(4) Produces little or no waste material.

人类活动圈中的绿色化学

(1) 在工业生态学框架内实践化学科学和制造。

(2) 可持续、安全且无污染。

(3) 消耗最少的材料和能源。

(4) 几乎不产生废料。

重要名词和术语解析

Technology: Technology refers to the ways in which humans do and make things with materials and energy.

Pollutant: A reasonable definition of a pollutant is a substance present in greater than natural concentration as a result of human activity that has a net detrimental effect upon its environment or upon something of value in that environment.

Contaminant: Contaminants, which are not classified as pollutants unless they have some detrimental effect, cause deviations from the normal composition of an environment.

Source: Every pollutant originates from a **source**.

Receptor: The receptor is anything that is affected by the pollutant.

Sink: Eventually, if the pollutant is long-lived, it may be deposited in a **sink**, a long-time repository of the pollutant.

Hazardous waste: A simple definition of a **hazardous waste** is that it is a potentially dangerous substance that has been discarded, abandoned, neglected, released, or designated as a waste material, or is one that may interact with other substances to pose a threat.

技术：技术是指人类用物质和能量来做事和制造物品的方式。

污染物：对污染物的合理定义是由于人类活动而导致的浓度高于自然浓度的一种物质，对该物质的环境或该环境中的某些有价值的东西具有有害影响。

污染成分：会导致环境其偏离正常组成的成分，除非具有一定的有害作用，否则不归为污染物。

来源：每种污染物均来自一种**来源**。

受体：受体是受污染物影响的任何物质。

汇：最终，如果污染物是长寿命的，则可能会沉积在**汇**中，这是污染物的长期储存库。

危险废物：**危险废物**的简单定义是，它是已被废弃、丢弃、忽略、释放或指定为废物的潜在危险物质，或者是可能与其他物质相互作用构成威胁的物质。

1.9 Environmental Chemistry of the Biosphere 生物圈环境化学

知识框架

The sustainability of a thriving biosphere is the top priority in maintaining a healthy Earth.

(1) Production of biomass that is the base of the food chain of organisms and is a source of renewable materials and fuels.

(2) Fixing of carbon and solar energy by photosynthesis, removal of CO_2 from the atmosphere.

(3) Biodegradation and detoxication of wastes and pollutants.

(4) Weathering of rocks in the geosphere.

(5) Mediation of crucial biogeochemical cycles.

(6) Production of some pollutants and toxicants, such as acid mine water by action of bacteria on pyrite or methymercury compounds by anoxic bacteria metabolizing inorganic mercury.

(7) Nitrification and denitrification processes. Toxicological Chemistry and Biochemistry.

繁荣的生物圈的可持续性是维护地球健康的重中之重。

(1) 生产生物质，这是生物体食物链的基础，也是可再生材料和燃料的来源。

(2) 通过光合作用固定碳和太阳能，从大气中去除 CO_2。

(3) 废物和污染物的生物降解和解毒。

(4) 地球大气中的岩石风化。

(5) 关键生物地球化学循环的中介。

(6) 产生某些污染物和有毒物质，如酸性矿水，是由于细菌对黄铁矿或甲基汞化合物的作用，通过缺氧细菌代谢无机汞形成的。

(7) 硝化和反硝化过程。毒物化学和生物化学。

重要名词和术语解析

Biology: Biology is the science of life. It is based on biologically synthesized chemical species, many of which exist as large molecules called macromolecules.

Biosphere: The **biosphere** is the name given to that part of the environment consisting of organisms and living biological material.

生物学：生物学是生命科学。它基于生物合成的化学物质，其中许多以大分子的形式存在。

生物圈：**生物圈**是由生物和生物材料组成的那部分环境的名称。

Ecology: Ecology is the science that deals with the relationships between living organisms with their physical environment and with each other.

Ecosystem: An ecosystem consists of an assembly of mutually interacting organisms and their environment in which materials are interchanged in a largely cyclical manner.

Habitat: The environment in which a particular organism lives is called its **habitat**.

Niche: The role of an organism in a habitat is called its **niche**.

Terrestrial environment: The terrestrial environment is based on land and consists of **biomes**, such as grasslands, savannas, deserts, or one of several kinds of forests.

Marine environment: The oceanic marine environment is characterized by saltwater and may be divided broadly into the shallow waters of the continental shelf composing the **neritic zone** and the deeper waters of the ocean that constitute the **oceanic region**.

Symbiotic environment: An environment in which two or more kinds of organisms exist together to their mutual benefit is termed a **symbiotic environment**.

Population: A particularly important factor in describing ecosystems is that of populations consisting of numbers of a specific species occupying a specific habitat.

Population explosion: Populations may be stable, or they may grow exponentially as a population explosion.

Population crash: A population explosion that is unchecked results in resource depletion, waste

生态学：生态学是处理生物体与其物理环境以及彼此之间关系的科学。

生态系统：生态系统由相互作用的生物及其环境组成，在该环境中，材料很大程度上以周期性的方式交换。

栖息地：一种特定生物所居住的环境称为其**栖息地**。

生态位：生物在栖息地中的作用称为**生态位**。

陆地环境：陆地环境以土地为基础，由**生物群落**组成，例如草原、热带稀树草原、沙漠或几种森林中的一种。

海洋环境：海洋环境以咸水为特征，可大致分为组成浅海区的大陆架**浅水区**和构成海洋区的海洋**深水区**。

共生环境：两种或两种以上的生物体共同生存以达到共同利益的环境称为**共生环境**。

种群：描述生态系统的一个特别重要的因素是由居住在特定栖息地的多种特定物种组成的种群。

种群激增：种群可能稳定，或者以种群激增的形式呈指数增长。

种群崩溃：未经控制的种群爆炸会导致资源枯竭、废物堆积

accumulation, and predation culminating in an abrupt decline called a **population crash**.

Behavior: Behavior in areas such as hierarchies, territoriality, social stress, and feeding patterns plays a strong role in determining the fates of populations.

和捕食,最终导致急剧下降,称为**种群崩溃**。

行为:在等级、领土、社会压力和进食方式等方面的行为在决定种群命运方面起着重要作用。

Exercises
习题

1-1 Much of what is known about Earth's past history is based upon paleo-environmental studies. How can past climatic conditions, temperature, and atmospheric carbon dioxide levels be inferred going back hundreds of thousands of years based on ice cores and even millions of years based on fossils?

1-2 The idea of climate change caused by human activities appears to be relatively recent. However, it was proposed quite some time ago in a paper entitled "On the Influence of Carbonic Acid in the Air upon the Temperature of the Ground." When was this paper published and who was the author? What were his credentials and credibility?

1-3 What would be the sources of the gaseous nitrogen oxides, NO and NO_2? Which secondary air pollutant would they form interacting with volatile hydrocarbons in the sunlight? Could acid rain result from these oxides and, if so, what would be the formula of the acid?

1-4 In the late 1800s there was concern that

1-1 关于地球过去的历史,许多已知的事都是根据古环境研究得出的。如何通过追溯数十万年前的冰芯甚至数百万年前的化石,推断出过去的气候条件、温度和大气中二氧化碳的水平?

1-2 由人类活动引起的气候变化的想法似乎是相对较新的。但是,它是在很久以前的一篇题为"关于空气中的碳酸对地面温度的影响"的论文中提出的。该论文何时出版,作者是谁,他的资历和信誉是什么?

1-3 气态氮氧化物 NO 和 NO_2 的来源是什么,它们会与阳光中的挥发性碳氢化合物相互作用而形成哪种二次空气污染物,这些氧化物会引起酸雨吗,如果是的话,酸的分子式是什么?

1-4 在1800年代后期,人们关

within the nitrogen biogeochemical cycle, not enough of the atmosphere's inexhaustible store of nitrogen was being "fixed" to chemical forms that could be utilized by plants and that food shortages would result from a shortage of fixed nitrogen. What happened to change this situation? In what respect did this development save many lives and how did it also make possible the loss of millions of people in warfare after about 1900?

1-5 In what important, fundamental respect does the phosphorus cycle differ from the carbon, oxygen, and nitrogen cycles?

1-6 Most people are aware that atmospheric carbon dioxide contributes to global warming and climate change. In what respect, however, is the atmosphere's carbon dioxide part of Earth's natural capital, that is, where would we be without it? What crucial natural phenomenon causes a slight, but perceptible change in atmospheric carbon dioxide levels over the course of a year?

1-7 What command and control regulations have been implemented in limiting this source of pollution? What "end-of-pipe" measures have been used? Suggest how the practices of green chemistry might serve as alternatives to these measures?

1-8 As it applies to environmental processes and pollution, the term "sink" is sometimes used. Explain what is meant by a sink as it applies to environmental pollution. In what sense is Earth's ability

注的是，在氮生物地球化学循环中，没有将大气中不竭的氮存储"固定"为植物可以利用的化学形式，并且缺乏固定的氮将引起食物短缺。发生了什么改变了这种情况？这种发展在什么方面挽救了许多生命，又如何使大约1900年后的战争中数百万人丧生？

1-5 磷循环与碳循环、氧循环和氮循环在什么重要的根本方面不同？

1-6 大多数人都知道，大气中的二氧化碳会导致全球变暖和气候变化。但是，从什么角度看，大气中的二氧化碳是地球自然资本的一部分，也就是说，如果没有它，我们将何去何从，哪种关键的自然现象会导致一年中大气中二氧化碳的含量发生轻微但可察觉的变化？

1-7 在限制污染源方面实施了哪些命令和控制法规，使用了哪些"末端治理"措施，绿色化学实践如何替代这些措施？

1-8 因为其适用于环境过程和污染，因此有时使用"汇"一词，解释一下汇对环境污染的含义。从什么意义上说，地球的自然

to act as a sink part of its natural capital? Explain.

1-9 In dealing with pollution and the potential for pollution, three approaches are pollution prevention, end-of-pipe measures, and remediation. What do these terms mean in terms of pollution control? Which is the most desirable, and which is the least? Explain.

1-10 With respect to increased production of corn to provide fuel ethanol, it is stated in this chapter that, "Increased demand for fertilizer in the form of chemically combined nitrogen means that more ammonia is synthesized using atmospheric nitrogen and impacting the nitrogen cycle." With respect to which resource of Earth's capital is the synthetic production of nitrogen fertilizer as it is now done a problem, and in respect to which resource is it not a problem. Explain.

资本中充当汇部分的能力是什么？并展开说明。

1-9 在处理污染和潜在污染方面，有三种方法分别为污染预防、末端治理和补救。这些术语在污染控制方面意味着什么，哪个是最理想的，哪个是最不理想的？展开说明。

1-10 关于增加玉米的产量以提供燃料乙醇，本章指出："对化学结合氮形式的肥料需求的增加，意味着使用大气氮会合成更多的氨，并影响氮循环。"关于氮肥的合成生产，对地球资本的哪种资源现在已经成为问题，而对于哪种资源则不是问题，并展开说明。

扫描二维码查看
习题答案

2 Environmental Chemistry of the Atmosphere
大气圈环境化学

2.1 The Atmosphere and Atmospheric Chemistry
大气与大气化学

2.1.1 The Atmospheric Composition and the Main Pollutants
大气组成与主要污染物

知识框架

(1) The atmosphere consists of the following (on a dry basis):
1) 78.1% N_2.
2) 21.0% O_2.
3) 0.9% Ar.
4) 0.04% CO_2.
5) Low levels of noble gas helium, neon, krypton, xenon.
6) Trace gases.

(2) Most of the mass of the atmosphere is very close to Earth's surface relative to Earth's diameter.

If earth were a classroom globe, virtually all air would be in a layer the thickness of the coat of varnish on the globe.

Photochemistry and Some Important Terms

(1) Photochemical reactions occur in the atmosphere when molecules absorb energy in the form of photons.

(1) 大气由以下物质组成（干重）：
1) 78.1% N_2；
2) 21.0% O_2；
3) 0.9% Ar；
4) 0.04% CO_2；
5) 稀有气体氦、氖、氪、氙的含量低；
6) 微量气体。

(2) 相对于地球直径，大部分大气都非常靠近地球表面。

如果地球是教室里的地球仪，那么实际上所有的空气都存在于与地球仪上清漆厚度相当的那一层中。

光化学和一些重要术语

(1) 当分子吸收光子形式的能量时，在大气中发生光化学反应。

1) Mostly in ultraviolet region of spectrum.

2) $E = h\nu$.

(2) A chemical species in an excited (energized) state is designated with an asterisk, *.

(3) The photochemical reaction of stratospheric ozone:

1) $O_3 + h\nu\ (\lambda < 420\text{nm}) \rightarrow O^* + O_2$.

2) The O_3 absorbs a photon of energy $h\nu$.

3) The O_3 undergoes photodissociation.

4) The oxygen atom product is excited denoted O^*.

(4) Free radicals with unpaired electrons shown with a dot:

$$H_2O_2 + h\nu \longrightarrow HO\cdot + HO\cdot$$

(5) Energy-absorbing third body, M, usually N_2 molecule:

$$O + O_2 + M \longrightarrow O_3 + M$$

Gaseous Oxides in the Atmosphere

(1) Low and variable levels of C, N, and S oxides.

Pollutants at elevated levels.

(2) Atmospheric methane.

1) From anoxic bacteria and underground sources.

2) Significant greenhouse gas.

3) Influences levels of hydroxyl radical (HO·), ozone, stratospheric water vapor.

(3) Hydrocarbons and photochemical smog.

Hydrocarbons required for photochemical smog formation.

1) 主要在光谱的紫外线区域。

2) $E = h\nu$。

(2) 处于激发（高能）状态的化学物质带有星号*。

(3) 平流层臭氧的光化学反应：

1) $O_3 + h\nu\ (\lambda < 420\text{nm}) \rightarrow O^* + O_2$；

2) O_3 吸收光子能量 $h\nu$；

3) O_3 发生光离解；

4) 氧原子产物被激发，表示为 O^*。

(4) 带有未配对电子的自由基带有一个点：

(5) 吸收能量的第三体 M，通常为 N_2 分子：

大气中的气态氧化物

(1) 低且可变含量的 C、N 和 S 氧化物。

污染物浓度越来越高。

(2) 大气甲烷。

1) 来自缺氧细菌和地下来源。

2) 重要的温室气体。

3) 影响羟基自由基 (HO·)、臭氧、平流层水蒸气的水平。

(3) 碳氢化合物和光化学烟雾。

光化学烟雾形成需要碳氢化合物。

(4) Particulate matter.
(5) Primary pollutants emitted directly.
(6) Secondary pollutants formed from reactions of primary pollutants.

（4）颗粒物。
（5）直接排放的一次污染物。
（6）由一次污染物反应形成的二次污染物。

重要名词和术语解析

Photochemical reaction：The most significant feature of atmospheric chemistry is the occurrence of photochemical reactions resulting from the absorption by molecules of **photons**.

Particle：Particles ranging from aggregates of a few molecules to pieces of dust readily visible to the naked eye are commonly found in the atmosphere.

Aerosol：Colloidal-sized particles in the atmosphere are called aerosols.

Primary pollutant：Primary pollutants in the atmosphere are those that are emitted directly.

Secondary pollutant：Secondary pollutants are formed by atmospheric chemical processes acting on primary pollutants and even nonpollutant species in the atmosphere.

光化学反应：大气化学的最显著特征是由于**光子**分子吸收而发生的光化学反应。

颗粒物：从几个分子的聚集体到肉眼容易看到的尘埃的颗粒物通常在大气中被发现。

气溶胶：大气中胶体大小的颗粒称为气溶胶。

一次污染物：大气中的一次污染物是直接排放的污染物。

二次污染物：二次污染物是由大气化学过程作用于大气中的一次污染物甚至非污染物质而形成的。

2.1.2 How the Atmosphere Got That Way and its Natural Capital
大气如何发展及其自然资本

知识框架

Importance of the Atmosphere

(1) Chemical and Biochemical Processes in Evolution of the Atmosphere.

1) Atmosphere's O_2 from cyanobacteria photosynthesis.

2) Photochemical formation of stratospheric ozone.

(2) Protective function of the atmosphere.

大气的重要性

（1）大气演化中的化学和生化过程。

1）蓝细菌光合作用产生的大气中的 O_2。

2）平流层臭氧的光化学形成。

（2）大气的保护功能。

1) Filters out harmful radiation.

2) Stabilizes temperature.

(3) Part of hydrologic cycle.

(4) Source of CO_2 for plant photosynthesis.

(5) Source of N for plant growth, industrial chemicals.

(6) Self-Purification of the Atmosphere.

1) Conversion of pollutants to harmless substances.

2) Dry deposition of contaminants.

3) Wet deposition of water soluble contaminants.

1) 过滤掉有害辐射。

2) 稳定温度。

(3) 水文循环的一部分。

(4) 植物光合作用的 CO_2 来源。

(5) 植物生长、工业化学品的氮源。

(6) 大气的自我净化。

1) 将污染物转化为无害物质。

2) 污染物的干沉降。

3) 水溶性污染物的湿沉降。

2.1.3 Physical Characteristics of the Atmosphere
大气的物理特征

重要名词和术语解析

Atmospheric science: Atmospheric science deals with the movement of air masses in the atmosphere, atmospheric heat balance, and atmospheric chemical composition and reactions.

Troposphere: The lowest layer of the atmosphere extending from sea level to an altitude of 10~16km is the **troposphere**, characterized by a generally homogeneous composition of major gases other than water and decreasing temperature with increasing altitude from the heat-radiating surface of the earth.

Adiabatic lapse rate: The extent of the temperature decrease for dry air with increasing altitude is known as the adiabatic lapse rate, which has a value of 9.8K/km.

Tropopause: The very cold temperature of the

大气科学：大气科学涉及空气在大气中的运动、大气热平衡以及大气化学成分和反应。

对流层：对流层是从海平面一直延伸到10~16km高度的**最底层大气**，其特征是除水以外的主要气体的成分基本均匀，并且温度随着离地球热辐射表面高度的升高而降低。

绝热递减率：干燥空气温度随高度升高而下降的程度称为绝热递减率，其值为9.8K/km。

对流层顶：对流层顶的极冷温

tropopause layer at the top of the troposphere serves as a barrier that causes water vapor to condense to ice so that it cannot reach altitudes at which it would photodissociate through the action of intense high-energy ultraviolet radiation.

Stratosphere: The atmospheric layer directly above the troposphere is the stratosphere, in which the temperature rises to a maximum of about −2℃ with increasing altitude.

Mesosphere: The absence of high levels of radiation-absorbing species in the mesosphere immediately above the stratosphere results in a further temperature decrease to about −92℃ at an altitude around 85km.

Thermosphere: Extending to the far outer reaches of the atmosphere is the thermosphere, in which the highly rarified gas reaches temperatures as high as 1200℃ by the absorption of very energetic radiation of wavelengths less than approximately 200nm by gas species in this region.

度起屏障作用，使水蒸气凝结成冰，使其无法到达会被强烈的高能紫外线辐射而被光解离的高度。

平流层：对流层正上方的大气层是平流层，随着海拔的升高，温度升高到最高约-2℃。

中间层：平流层正上方的中间层缺乏高浓度的吸收辐射的物质，导致在85km左右的高度温度进一步下降至-92℃。

热层：热层延伸到大气的最外层，在该层中，极度稀有的气体通过吸收该区域中气体发出的波长小于约200nm的高能辐射而达到高达1200℃的温度。

2.1.4 Energy Transfer in the Atmosphere 大气中的能量转移

重要名词和术语解析

Solar constant: The solar energy flux reaching the atmosphere is huge, amounting to 1.34×10^3 watts per square meter (19.2 kcal❶ per minute per square meter) perpendicular to the line of solar flux at the top

太阳常数：到达大气层的太阳能通量很大，垂直于大气层顶部的太阳通量线等于每平方米 1.34×10^3 W（每平方米每分钟19.2kcal❶）。这个数值是太阳常数，也可以称为

---

❶ 1kcal=4.1868kJ。

of the atmosphere. This value is the solar constant, and may be termed insolation, which stands for "incoming solar radiation".

Conduction: Conduction of energy occurs through the interaction of adjacent atoms or molecules without the bulk movement of matter and is a relatively slow means of transferring energy in the atmosphere.

Convection: Convection involves the movement of whole masses of air, which may be either relatively warm or cold.

Sensible heat: As well as carrying sensible heat due to the kinetic energy of molecules, convection carries **latent heat** in the form of water vapor which releases heat as it condenses.

Radiation: Radiation of energy in earth's atmosphere occurs through electromagnetic radiation in the infrared region of the spectrum.

Albedo: An important aspect of solar radiation that reaches earth's surface is the percentage reflected from the surface, described as albedo.

传导：能量传导是通过相邻原子或分子的相互作用而发生的，而没有物质的大量运动，是在大气中传递能量的相对较慢的方式。

对流：对流涉及整个空气的运动，可能相对温暖或较冷。

显热：由于分子的动能也会产生显热，对流还会以水蒸气的形式携带潜热，并在冷凝时释放热量。

辐射：地球大气中的能量辐射是通过光谱的红外区域中的电磁辐射发生的。

反照率：到达地球表面的太阳辐射的一个重要方面是从表面反射的百分比，称为反照率。

2.1.5 Atmospheric Mass Transfer, Meteorology and Weather 大气传质、气象和天气

知识框架

(1) Meteorology is the science of physical atmospheric phenomena.

(2) Weather: Short-term variations in:

1) Temperature.
2) Clouds.
3) Wind.

(1) 气象学是物理大气现象的科学。

(2) 天气：影响天气短期变化的有：

1) 温度；
2) 云；
3) 风；

4) Humidity.

5) Pressure.

6) Horizontal visibility.

7) Precipitation type and quantity.

(3) Climate: Long-term weather conditions.

(4) Atmospheric Water in Energy and Mass Transfer.

Carries energy as latent heat released when water vapor condenses.

(5) Humidity is water content of air.

Relative humidity is % saturation level.

(6) Water condenses below dew point. Condensation nuclei.

(7) Clouds are composed of microdroplets of water.

Coalesce to form larger droplets and precipitation.

Distinct air masses in the atmosphere

(1) Uniform temperature and water vapor content.

(2) Horizontally homogeneous.

(3) Conditions and movement affect pollutant reactions, effects, and dispersal.

(4) Air masses separated by fronts.

1) Warm fronts.

2) Cold fronts.

(5) Wind is horizontally moving air.

(6) Air currents are vertically moving air.

Topographical effects

Topography, surface configuration and relief

4) 湿度；

5) 压力；

6) 水平能见度；

7) 降水类型和量。

(3) 气候：长期天气条件。

(4) 能量和质量传递中的大气水。

以潜热形式携带能量，当水蒸气冷凝时释放能量。

(5) 湿度是空气中的水分。相对湿度是指饱和的程度（%）。

(6) 水在露点以下凝结。凝聚核。

(7) 云由微滴水组成。

聚结形成更大的液滴和沉淀。

大气中不同的气团

(1) 温度和水蒸气含量均匀。

(2) 水平均质。

(3) 条件和运动会影响污染物的反应、影响和扩散。

(4) 气团被锋面隔开。

1) 暖锋。

2) 冷锋。

(5) 风在水平移动空气。

(6) 气流是垂直移动的空气。

地形影响

地形、表面构造和起伏特

features strongly affect winds and air currents.

Cyclonic storms:

(1) Hurricanes (Atlantic).

(2) Typhoons (Pacific).

征强烈影响风和气流。

气旋风暴:

(1)飓风(大西洋);

(2)台风(太平洋)。

重要名词和术语解析

Meteorology: Meteorology is the science of atmospheric phenomena, encompassing the study of the movement of air masses as well as physical forces in the atmosphere—heat, wind, and transitions of water, primarily liquid to vapor or vice versa.

Weather: Short-term variations in the state of the atmosphere constitute weather.

Climate: Longer-term variations and trends within a particular geographical region in those factors that compose weather are described as climate.

Humidity: The water vapor content of air can be expressed as humidity.

Relative humidity: expressed as a percentage, describes the amount of water vapor in the air as a ratio of the maximum amount that the air can hold at that temperature.

Dew point: Air with a given relative humidity can undergo any of several processes to reach the saturation point at which water vapor condenses in the form of rain or snow. For this condensation to happen, air must be cooled below a temperature called the dew point, and **condensation nuclei** must be present.

Front: Solar energy received by earth is

气象学:气象学是大气现象的科学,涵盖了空气质量运动以及大气中的物理力的研究,包括热、风和水的转换,主要是液体到蒸气的转化,反之亦然。

天气:大气状态的短期变化构成天气。

气候:特定地理区域内构成天气的那些因素的长期变化和趋势称为气候。

湿度:空气中的水蒸气含量可以表示为湿度。

相对湿度:用百分比表示,表示空气中水蒸气的量与空气在该温度下可容纳的最大量之比。

露点:具有给定相对湿度的空气可以经历多个过程中的任何一个过程,以达到饱和点,在该饱和点,水蒸气以雨或雪的形式冷凝。为了使这种冷凝发生,必须将空气冷却到称为露点的温度以下,并且必须存在**冷凝核**。

锋面:地球接收的太阳能主要

largely redistributed by the movement of huge masses of air with different pressures, temperatures, and moisture contents separated by boundaries called fronts.

Wind: Horizontally moving air is called wind, whereas vertically moving air is referred to as an **air current**.

Dry adiabatic lapse rate: If there is no condensation of moisture from the air, the cooling effect is about 10℃ per 1000 meters of altitude, a figure known as the dry adiabatic lapse rate.

Moist adiabatic lapse rate: This partially counteracts the cooling effect of the expanding air, giving a moist adiabatic lapse rate of about 6℃/1000m.

Convection current: Air currents are largely convection currents formed by differential heating of air masses.

Topography: the surface configuration and relief features of the earth's surface may strongly affect winds and air currents.

Local convective wind: Differential heating and cooling of land surfaces and bodies of water can result in local convective winds, including land breezes and sea breezes at different times of the day along the seashore, as well as breezes associated with large bodies of water inland.

Convection column: The warm, moist mass of air thus produced moves from a region of high pressure to one of low pressure, and cools by expansion as it rises in what is called a convection column.

通过具有不同压力、温度和水分含量的大量空气的移动来重新分配，这些空气被称为锋面的边界分隔开。

风：水平移动的空气称为风，而垂直移动的空气称为**气流**。

干绝热递减率：如果空气中没有水分凝结，则冷却效果约为每1000m高度10℃，该数字称为干绝热递减率。

湿绝热递减率：这部分抵消了膨胀空气的冷却效果，使湿绝热递减率约为6℃/1000m。

对流：气流主要是由不同热量的气团形成的对流。

地形：地球的表面构造和起伏特征可能会强烈影响风和气流。

局部对流风：陆地表面和水体的不同加热和冷却会导致局部对流风，包括一天中不同时间沿海岸的陆风和海风，以及与内陆大型水体有关的微风。

对流柱：如此产生的温暖潮湿的空气从高压区域移至低压区域，随着在对流柱中升高因膨胀而冷却。

Coriolis effect: Coriolis effect results in spiral-shaped air circulation patterns called cyclonic or anticyclonic, depending upon the direction of rotation.

科里奥利效应：根据旋转方向科里奥利效应会形成称为气旋或反气旋的螺旋形空气循环模式。

2.1.6 Inversions and Air Pollution
逆温与空气污染

知识框架

（1）Radiation inversions are likely to form in still air at night when the Earth is no longer receiving solar radiation.

（2）Subsidence inversions, often accompanied by radiation inversions, can become very widespread.

（3）A marine inversion is produced during the summer months when cool air laden with moisture from the ocean blows onshore and under warm, dry inland air.

（1）当地球不再接收太阳辐射时，夜晚的静止空气中可能会形成辐射逆温。

（2）沉降逆温通常伴随着辐射逆温会变得非常普遍。

（3）夏季，当海洋中充满湿气的凉爽空气吹向陆上，位于温暖干燥的内陆空气之下时，就会产生海洋逆温。

重要名词和术语解析

Temperature inversion: Temperature inversions can be caused in several ways in which higher temperatures with increasing altitude limit vertical circulation of air and cause air stagnation and trapping of air pollutants near Earths surface.

逆温：逆温可以通过多种方式引起，随着海拔的升高而升高的温度会限制空气的垂直循环，并导致空气停滞和空气污染物被困在地球表面附近。

2.1.7 Global Climate and Microclimate
全球气候与微气候

知识框架

Climate
（1）Characteristic of a particular region.
（2）Varies with season.

气候
（1）特定区域的特征。
（2）因季节而异。

Example: Alternating monsoons and dry seasons.

(3) Ice age manifested by long-term change in climate.

Humans may be modifying climate largely by pumping carbon dioxide into the atmosphere causing warming.

Microclimate
Highly localized climate.
Example: At soil surface shaded by plants.

Effects of urbanization on microclimate
(1) Heat dome over cities.
(2) City atmosphere up to 5℃ warmer.
(3) Counteracting cooling effect from particulate matter.

例：季风和旱季交替。

（3）冰期以气候的长期变化为特征。

人类可能通过将二氧化碳泵入大气而引起变暖，从而在很大程度上改变了气候。

微气候
高度局部化的气候。
例：在被植物遮蔽的土壤表面。

城市化对微气候的影响
（1）城市上空的热圆顶。
（2）城市大气温度升高5℃。
（3）抵消颗粒物的冷却效果。

重要名词和术语解析

Climate: Perhaps the single most important influence on Earth's environment is climate, consisting of long-term weather patterns over large geographical areas.

Monsoon: seasonal variations in wind patterns between oceans and continents.

Microclimate: Highly localized climatic conditions are termed the microclimate.

Dew: These lower temperatures result in condensation of dew on vegetation and the soil surface, thus providing a relatively more moist microclimate near ground level.

Heat dome: a city is capped by a heat dome

气候：对地球环境的影响最重要的因素可能是气候，包括大范围地理区域的长期天气模式。

季风：大洋和大洲之间风的季节性变化。

微气候：高度局部化的气候条件称为微气候。

露水：这些较低的温度导致露水凝结在植被和土壤表面上，从而在地面附近提供相对较湿的微气候。

热圆顶：城市被热圆顶所覆盖，

in which the temperature is as much as 5℃ warmer than in the surrounding rural areas, such that large cities have been described as "heat islands".

其温度比周围的农村地区高5℃，因此大城市被称为"热岛"。

2.1.8 Chemical and Photochemical Reactions 化学和光化学反应

知识框架

Study of atmospheric chemistry is complicated

(1) Effects of photochemical energy input.

(2) Extreme dilution of species in air.

(3) Container walls complicate laboratory study.

大气化学的研究很复杂

(1) 光化学能输入的影响。

(2) 空气中物种极度稀释。

(3) 容器壁使实验室研究复杂化。

Major categories of atmospheric chemical species

(1) Inorganic oxides (CO, CO_2, NO_2, SO_2).

(2) Oxidants (O_3, H_2O_2, $HO·$, $HO_2·$, and $ROO·$ radicals, NO_3).

(3) Reductants (CO, SO_2, H_2S).

(4) Organics (such as CH_4, most also reductants).

(5) Oxidized organics (carbonyls, organic nitrates).

(6) Photochemically active species (NO, formaldehyde).

(7) Acids (H_2SO_4), bases (NH_3), salts (NH_4HSO_4).

大气化学物质的主要类别

(1) 无机氧化物（CO，CO_2，NO_2，SO_2）。

(2) 氧化剂（O_3，H_2O_2，$HO·$，$HO_2·$，和 $ROO·$ 自由基，NO_3）。

(3) 还原剂（CO，SO_2，H_2S）。

(4) 有机物（如 CH_4，大多数也是还原剂）。

(5) 氧化有机物（羰基，有机硝酸盐）。

(6) 光化学活性物质（NO，甲醛）。

(7) 酸（H_2SO_4），碱（NH_3），盐（NH_4HSO_4）。

(8) Unstable reactive species (NO_2^*, $HO\cdot$).

Photochemical Processes

(1) Initiated when a molecule absorbs a photon of electromagnetic radiation to produce an excited species, *.

1) $NO_2 + h\nu \rightarrow NO_2^*$.

2) Usually in ultraviolet region.

(2) Loss of excess energy from an excited state may occur by several processes including:

1) Dissociation: $NO_2^* \rightarrow NO + O$.

2) Luminescence: $NO_2^* \rightarrow NO_2 + h\nu$.

3) Photoionization: $N_2^* \rightarrow N_2^+ + e^-$.

Ions in the Atmosphere

(1) Ionosphere above about 50km.

1) From photoionization by solar ultraviolet.

2) Raises at night as ions recombine.

(2) Plasmasphere just beyond the ionosphere consisting of electrons and positively charged molecular and radical fragments formed by interaction of solar wind and Earth's magnetosphere.

(3) Plasmasphere protects from potentially catastrophic solar radiation events. The "Great Solar Storm of 1859" seriously disrupted telegraph systems.

(4) Ions at Lower Altitudes in the Troposphere.

Free Radicals

(1) Reactive species with unpaired electrons

(8) 不稳定的反应物种 (NO_2^*，$HO\cdot$)。

光化学过程

(1) 当分子吸收电磁辐射的光子以产生受激物质"*"时引发。

1) $NO_2 + h\nu \rightarrow NO_2^*$。

2) 通常在紫外线区域。

(2) 激发态可能会通过以下几个过程造成多余能量的损失：

1) 离解：$NO_2^* \rightarrow NO + O$；

2) 发光：$NO_2^* \rightarrow NO_2 + h\nu$；

3) 光电离：$N_2^* \rightarrow N_2^+ + e^-$。

大气中的离子

(1) 电离层约50km。

1) 来自太阳紫外线的光电离。

2) 晚上随着离子复合而上升。

(2) 等离子体层位于电离层的正上方，由电子以及由太阳风和地球磁层相互作用形成的带正电的分子和自由基碎片组成。

(3) 等离子体层可防止潜在的灾难性太阳辐射事件。1859年的大太阳风暴严重破坏了电报系统。

(4) 对流层中较低海拔的离子。

自由基

(1) 具有未配对电子的反

(2) Generally formed by photochemical reactions or reactions of molecules with other free radicals:

$$H_3C-\overset{\overset{O}{\|}}{C}-H + h\nu \longrightarrow H_3C\cdot + \cdot\overset{\overset{O}{\|}}{C}-H$$

(3) Two free radicals may react to form a stable species.

(4) Hydroxyl radical in the atmosphere.

1) HO· is the most important free radical in the atmosphere.

2) Produced by many reactions such as:

$CH_4 + O$ (from photodissociation of NO_2, 来自 NO_2 的光解离) $\longrightarrow H_3C\cdot + HO\cdot$

3) Removed by many reactions, especially with CO or CH_4.

(5) Hydroperoxyl radical in the atmosphere:

1) HOO·.

2) Less important than HO·, but still significant.

Evolution of the Atmosphere

Atmospheric oxygen from photosynthesis:

(1) $CO_2 + H_2O + h\nu \rightarrow \{CH_2O\} + O_2$.

(2) Evidence from iron oxide deposits.

(2) 通常由光化学反应或分子与其他自由基的反应形成:

(3) 两个自由基可以反应形成稳定的物种。

(4) 大气中的羟基自由基。

1) HO·是大气中最重要的自由基。

2) 由许多反应产生,如:

3) 通过许多反应去除,尤其是与 CO 或 CH_4 反应。

(5) 大气中的过氧羟基自由基:

1) HOO·;

2) 不如 HO·重要,但仍然重要。

大气的演变

光合作用产生的大气氧:

(1) $CO_2 + H_2O + h\nu \rightarrow \{CH_2O\} + O_2$;

(2) 来自氧化铁沉积物的证据。

重要名词和术语解析

Photochemical reaction: The absorption by chemical species of light, broadly defined here to include ultraviolet radiation from the sun, can bring about reactions, called photochemical reactions, which do not otherwise occur under the conditions (particularly the temperature)

光化学反应:化学物质吸收光(广义定义为包括来自太阳的紫外线辐射)可以引起光化学反应,该反应没有光的同等条件下(尤其是温度)不会在介质中发生。

of the medium in the absence of light.

Electronically excited molecule: A species such as NO_2 may absorb light of energy $h\nu$, producing an electronically excited molecule, $NO_2 + h\nu \rightarrow NO_2^*$.
Designated in the reaction above by an asterisk, *.

Free radical: Atoms or molecular fragments with unshared electrons, called free radicals, and ions consisting of electrically-charged atoms or molecular fragments.

Excited singlet state: In some cases the electron thus promoted retains a spin opposite to that of its former partner, giving rise to an excited singlet state.

Excited triplet state: The spin of the promoted electron is reversed, such that it has the same spin as its former partner; this gives rise to an excited triplet state.

Quantum: Normally, the first step in a photochemical process is the activation of the molecule by the absorption of a single unit of photochemical energy characteristic of the frequency of the light called a quantum of light.

Physical quenching: Loss of energy to another molecule or atom (M) by physical quenching.

Dissociation: Dissociation of the excited molecule (the process responsible for the predominance of atomic oxygen in the upper atmosphere).

电子激发分子：NO_2 之类的物质可能吸收能量为 $h\nu$ 的光，从而产生电子激发分子，$NO_2 + h\nu \rightarrow NO_2^*$。
在上述反应中以星号"*"表示。

自由基：具有未共享电子的原子或分子片段，称为自由基，以及由带电原子或分子片段组成的离子。

激发单重态：在某些情况下，由此激发的电子保持与其之前的配对电子相反的自旋，从而产生激发单重态。

激发三重态：被激发的电子的自旋被反转，因此它具有与其之前的配对电子相同的自旋；这会导致激发三重态。

量子：通常情况下，光化学过程的第一步是通过吸收具有光频率特征的单个光化学能量，即称为光量子，来激活分子。

物理淬灭：通过物理淬灭使能量损失到另一个分子或原子（M）。

解离：受激分子的解离（导致高层大气中原子氧占优势的过程）

Luminescence: Luminescence consisting of loss of energy by the emission of electromagnetic radiation.

Fluorescence: If the re-emission of light is almost instantaneous, luminescence is called fluorescence, and if it is significantly delayed, the phenomenon is **phosphorescence**.

Chemiluminescence: Chemiluminescence is said to occur when the excited species is formed by a chemical process.

Intermolecular energy transfer: an excited species transfers energy to another species which then becomes excited. A subsequent reaction by the second species is called a **photosensitized** reaction.

Intramolecular transfer: Intramolecular transfer in which energy is transferred within a molecule.

Photoionization: Photoionization through loss of an electron.

Ionosphere: At altitudes of approximately 50km and up, ions are so prevalent that the region is called the ionosphere.

发光：发光是指由于电磁辐射的发射而导致的能量损失。

荧光：如果几乎瞬时发出光，则发光称为荧光，如果显著延迟，则现象为**磷光**。

化学发光：化学发光是通过化学过程形成激发物种时发生的。

分子间能量转移：激发态的物质将能量转移到另一种物质，使其被激发。第二种物质的后续反应称为**光敏**反应。

分子内转移：分子内转移，其中能量在分子内转移。

光电离：通过电子损失进行光电离。

电离层：在大约 50km 以上的高度，离子非常普遍，因此该区域称为电离层。

2.1.9 Acid-Base Reactions in the Atmosphere
大气中的酸碱反应
知识框架

(1) Rainwater weakly acidic from CO_2:

$$CO_2 + H_2O \longleftrightarrow H^+ + HCO_3^-$$

(2) Pollutant SO_2 is more acidic than CO_2.

(1) 雨水由于 CO_2 存在呈弱酸性：

(2) 污染物 SO_2 比 CO_2 酸性更高。

(3) Strong acid H_2SO_4, HNO_3, and HCl are responsible for damaging acid rain.

(3) 强酸 H_2SO_4、HNO_3 和 HCl 会产生破坏性酸雨。

2.1.10 Reactions of Atmospheric Oxygen 大气中的氧的反应

知识框架

(1) The resulting division of the atmosphere into a lower section with a uniform molecular weight and a higher region with a nonuniform molecular weight is the basis for classifying these two atmospheric regions as the homosphere and heterosphere, respectively.

(1) 最终将大气分为分子量均匀的下部区域，和分子量不均匀的上部区域，这是将这两个大气区域分别划分为均质层和异质层的基础。

(2) This emitted light is partially responsible for airglow, a very faint electromagnetic radiation continuously emitted by the earth's atmosphere.

(2) 发出的光是造成气辉的部分原因，气辉是地球大气层连续发出的非常微弱的电磁辐射。

2.1.11 Reactions of Atmospheric Nitrogen 大气中的氮的反应

知识框架

(1) N_2 molecule is very stable.

1) No significant tropospheric chemical or photochemical reactions of N_2.

2) N_2 is the most common energy-absorbing third body, "M", in atmospheric chemistry.

(2) Fixation of N from atmospheric N_2 is an important environmental process.

1) Biochemically by specialized bacteria.

2) Chemically by NH_3 synthesis.

Significant source of fixed N produced in the anthrosphere and released to the nitrogen cycle.

(1) N_2 分子非常稳定。

1) N_2 在对流层无明显的化学或光化学反应。

2) 氮是大气化学中最常见的吸收能量的第三体"M"。

(2) 从大气中固氮是一个重要的环境过程。

1) 由特殊细菌生化。

2) 化学合成 NH_3。

在人类活动圈产生大量固定氮并释放到氮循环中。

(3) N compounds such as NO and NO_2 are very active species in tropospheric chemistry.

（3）N 的化合物（如 NO 和 NO_2）在对流层化学中是非常活跃的物种。

2.1.12 Atmospheric Water
大气中的水

知识框架

(1) Normal range 1%~3% by volume.

(2) Vapor responsible for atmospheric temperature stability.

(3) Hydrologic cycle.

(4) Crucial in atmospheric energy transfer.

(5) Tropopause prevents water vapor transfer from troposphere to stratosphere.

(6) Stratospheric water from following several-step process:

$$CH_4 + 2O_2 + h\nu \longrightarrow CO_2 + 2H_2O$$

(7) Stratospheric water produces hydroxyl radical:

$$H_2O + h\nu \longrightarrow HO \cdot + H$$

（1）正常体积分数为 1%~3%。

（2）水蒸气对大气温度稳定性有贡献。

（3）水文循环。

（4）大气能量转移至关重要。

（5）对流顶层防止水蒸气从对流层转移到平流层。

（6）平流层水来自以下过程（多步反应）：

（7）平流层水产生羟基自由基：

2.1.13 Influence of the Anthrosphere
人类活动圈的影响

知识框架

(1) Many air pollutants from the anthrosphere:

1) Particles affecting visibility.
2) Acid-forming gases such as SO_2.
3) Nitrogen oxides and hydrocarbons forming photochemical smog.

(2) Two major kinds of species affecting global climate.

1) Chlorofluorocarbons that deplete stratospheric ozone.

（1）来自人类活动圈的许多空气污染物：

1) 影响可见度的颗粒；
2) 生成酸的气体，例如 SO_2；
3) 氮氧化物和碳氢化合物形成光化学烟雾。

（2）影响全球气候的两种主要物种。

1) 消耗平流层臭氧的氯氟烃。

2) Greenhouse gases that cause global warming.
①Primarily CO_2.
②Other gases such as CH_4.

2) 导致全球变暖的温室气体。
①主要是二氧化碳；
②其他气体，如 CH_4。

2.1.14 Chemical Fate and Transport in the Atmosphere
大气中的化学归宿和迁移

知识框架

(1) Considers the following regarding airborne pollutants:
1) Sources.
2) Transport.
3) Dispersal.
4) Fluxes.

(2) Atmosphere/surface boundary interaction:
1) Rock/soil.
2) Water.
3) Vegetation.

(3) Transport and dispersal:
1) Movement of air masses.
2) Diffusive and Fickian transport.

Long-range movement such as radionuclides from Chernobyl reactor meltdown.

Distillation of semivolatile organic pollutants to polar regions.

(1) 关于大气污染物考虑以下方面：
1) 来源；
2) 迁移；
3) 分散；
4) 通量。

(2) 大气/表面边界相互作用：
1) 岩石/土壤；
2) 水；
3) 植被。

(3) 迁移和分散：
1) 气团运动；
2) 扩散和菲克式迁移。

远程运动，如切尔诺贝利核反应堆熔化产生的放射性核素。

将半挥发性有机污染物蒸馏至极地。

2.2 Particles in the Atmosphere
大气颗粒物

2.2.1 Particles in the Atmosphere
大气中的粒子

知识框架

(1) Commonly called particulates.

(1) 通常称为颗粒。

(2) Size of about 0.5mm or less.

(3) Particles may be:

1) Inorganic.

2) Organic.

3) Biological (pollen, microorganisms).

(4) Primary particles emitted directly to the atmosphere:

1) Desert sand.

2) Sea salt.

3) Pollen.

(5) Secondary particles formed from gases in the atmosphere:

1) H_2SO_4.

2) Ammonium salts.

3) Photochemical smog.

(6) Particles undergo aging:

1) Gas exchange with air.

2) Water loss or uptake.

3) Chemical reactions within particles.

Terms pertaining to particles

(1) Aerosol: Colloidal-sized atmospheric particles.

(2) Condensation aerosol: Formed from gas or vapors.

(3) Dispersion aerosol: Formed from grinding bulk solids or dispersion of liquids.

(4) Fog: High level of water droplets.

(5) Haze: Decreased visibility due to particles.

(6) Mists: Liquid particles.

(7) Smoke: From incomplete combustion of carbonaceous fuel.

（2）尺寸约0.5mm或更小。

（3）颗粒可能是：

1）无机；

2）有机；

3）生物（花粉，微生物）。

（4）一次颗粒物直接排放到大气中：

1）沙漠沙；

2）海盐；

3）花粉。

（5）由大气中的气体形成的二次颗粒物：

1）H_2SO_4；

2）铵盐；

3）光化学烟雾。

（6）颗粒老化：

1）与空气交换气体；

2）水分流失或吸收；

3）颗粒内的化学反应。

有关颗粒物的术语

（1）气溶胶：胶体大小的大气颗粒。

（2）冷凝气溶胶：由气体或蒸气形成。

（3）分散气溶胶：由研磨分散的固体或液体分散形成。

（4）雾：高浓度的水滴。

（5）霾：由于颗粒而导致的可见度降低。

（6）薄雾：液体颗粒。

（7）烟：来自含碳燃料的不完全燃烧。

重要名词和术语解析

Particulate: **Particulates** is a term that has come to stand for particles in the atmosphere, although particulate matter or simply particles, is preferred usage.

Aerosol: Atmospheric **aerosols** are solid or liquid particles smaller than 100μm in diameter.

Mist: Liquid particulate matter, **mist**, includes raindrops, fog, and sulfuric acid mist.

颗粒物:颗粒物是代表大气中粒子的术语,更常用的说法是颗粒物质或简单称为粒子。

气溶胶:大气气溶胶是直径小于100μm的固体或液体颗粒。

薄雾:液体颗粒物,薄雾,包括雨滴、雾和硫酸雾。

2.2.2 Physical Behavior of Atmospheric Particles
大气颗粒的物理行为

知识框架

Stokes' law for particles of intermediate size:

$$v = \frac{gd^2(\rho_1 - \rho_2)}{18\eta}$$

where, v——settling velocity,
g——acceleration of gravity,
r_1——particle density,
r_2——air density,
h_2——air viscosity.

中等尺寸粒子的斯托克斯定律为:

$$v = \frac{gd^2(\rho_1 - \rho_2)}{18\eta}$$

式中 v——沉降速度;
g——重力加速度;
r_1——粒子密度;
r_2——空气密度;
h_2——空气黏度。

重要名词和术语解析

Particle size: **Particle size** usually expresses the diameter of a particle, though sometimes it is used to denote the radius.

Stokes diameter: Stokes' law can also be used to express the effective diameter of an irregular

粒径:粒径有时表示颗粒的直径,尽管有时用它来表示半径。

斯托克斯直径:斯托克斯定律也可以用来表示不规则非球形

nonspherical particle. These are called **Stokes diameters** (aerodynamic diameters) and are normally the ones given when particle diameters are expressed.

Mass median diameter: Mass median diameter (MMD) may be used to describe aerodynamically equivalent spheres having an assigned density of $1g/cm^3$ at a 50% mass collection efficiency, as determined in sampling devices calibrated with spherical aerosol particles having a known uniform size.

Brownian motion: Extremely small particles are subject to **Brownian motion** resulting from random movement due to collisions with air molecules and do not obey Stokes' Law.

粒子的有效直径,称为**斯托克斯直径**(空气动力学直径),通常是表示颗粒直径时给出的直径。

质量中值直径:质量中值直径(MMD)可用于描述在50%的质量收集效率下分配密度为$1g/cm^3$的空气动力学等效球体,是在用已知均一尺寸球形气溶胶颗粒校准的采样设备中确定的。

布朗运动:由于与空气分子的碰撞而导致的随机运动会导致极小的粒子发生**布朗运动**,因此不遵守斯托克斯定律。

2.2.3 Physical Processes for Particle Formation 粒子形成的物理过程

知识框架

(1) Physical processes produce dispersion aerosols:
1) Generally larger, above $1\mu m$.
2) Less respirable and less harmful than condensation aerosols.
(2) Natural sources include:
1) Sea spray.
2) Windblown dust.
3) Volcanic dust.
(3) Sources from human activities include:
1) Rock quarries.
2) Disturbed soil.
3) Dust from crop harvesting.

(1) 物理过程产生分散气溶胶:
1) 通常较大,大于$1\mu m$;
2) 比冷凝气溶胶的呼吸性和危害性小。
(2) 天然来源包括:
1) 海浪喷雾;
2) 飞尘;
3) 火山灰。
(3) 人类活动的来源包括:
1) 采石场;
2) 受干扰的土壤;
3) 作物收割产生的粉尘。

重要名词和术语解析

Dispersion aerosol: Dispersion aerosols, such as dusts, formed from the disintegration of larger particles are usually above 1μm in size.

分散气溶胶：较大颗粒崩解而形成的**分散气溶胶**（如粉尘）通常大于 1μm。

2.2.4 Chemical Processes for Particle Formation 颗粒形成的化学过程

知识框架

Particles from chemical processes

(1) Generally smaller, below 1μm.
(2) More respirable.
(3) Higher organic contents.
(4) Higher contents of toxic substances.
1) Toxic elements (arsenic).
2) Carcinogenic organics.

Inorganic Particles Formation

(1) Oxides: $3FeS_2+8O_2 \rightarrow Fe_3O_4+6SO_2$.

(2) Sulfuric acid: $2SO_2 + O_2 + 2H_2O \rightarrow 2H_2SO_4$.

(3) Salts: H_2SO_4 (droplet) $+ 2NH_3 \rightarrow (NH_4)_2SO_4$ (particulate).

Example of sodium sulfate forming on NaCl particles (from sea salt):

$$2NaCl+2HO\cdot \longrightarrow 2NaOH+Cl_2$$
$$Cl_2+2NaOH \longrightarrow NaOCl+NaCl+H_2O$$
$$2NaOH+H_2SO_4 \longrightarrow Na_2SO_4+H_2O$$
$$SO_2+2NaOH+\frac{1}{2}O_2 \longrightarrow Na_2SO_4+H_2O$$

化学过程中产生的颗粒

(1) 通常较小，小于 1μm。
(2) 更易吸入。
(3) 有机含量更高。
(4) 有毒物质含量较高。
1) 有毒元素（砷）；
2) 致癌有机物。

无机颗粒形成

(1) 氧化物：$3FeS_2+8O_2 \rightarrow Fe_3O_4+6SO_2$。

(2) 硫酸：$2SO_2 + O_2 + 2H_2O \rightarrow 2H_2SO_4$。

(3) 盐：H_2SO_4（液滴）$+ 2NH_3 \rightarrow (NH_4)_2SO_4$（颗粒）。

例如：在 NaCl 颗粒（来自海盐）上形成硫酸钠中：

Organic Particle Formation

Polycyclic Aromatic Hydrocarbon (PAH) Synthesis:

有机颗粒的形成

多环芳烃（PAH）合成：

重要名词和术语解析

Pyrosynthesis: Low-molar-mass hydrocarbons form PAHs by **pyrosynthesis**. This happens at temperatures exceeding approximately 500℃ at which carbon-hydrogen and carbon-carbon bonds are broken to form free radicals.

Pyrolysis: Polycyclic aromatic compounds may be formed from higher alkanes present in fuels and plant materials by the process of **pyrolysis**, the "cracking" of organic com-pounds to form smaller and less stable molecules and radicals.

<u>热合成</u>：低摩尔质量烃通过**热合成**形成多环芳烃。这发生在超过约500℃的温度下，在该温度下碳氢键和碳碳键断裂形成自由基。

<u>热解</u>：多环芳族化合物可以通过**热解**过程由燃料和植物材料中存在的高级烷烃形成，有机化合物"裂解"形成较小且不稳定的分子和自由基。

2.2.5 The Composition of Inorganic Particles 无机颗粒的组成

知识框架

(1) Sources of Elements in Inorganic Particles.

1) Al, Fe, Ca, Si: Soil erosion, rock dust, coal combustion.

2) C: Incomplete combustion of carbonaceous fuels.

（1）无机颗粒中元素的来源。

1）水土流失、岩尘、煤炭燃烧。

2）含碳燃料的不完全燃烧。

3) Na, Cl: Marine aerosols, organohalide polymer burning.

4) Sb, Se: Very volatile, combustion of oil, coal.

5) V: Combustion of residual petroleum.

6) Zn: In small particles, from combustion.

7) Pb: Combustion of fuels and wastes containing lead, less now with phaseout of leaded gasoline.

(2) Fly Ash.

Residue from fossil fuel combustion.

(3) Asbestos.

1) Formerly a problem from sources such as wear on brake linings.

2) Now phased out of general use.

3) 海洋气溶胶、有机卤化物聚合物燃烧。

4) 极易挥发，石油、煤炭燃烧。

5) 残留石油燃烧。

6) 在小颗粒中，来自燃烧。

7) 含铅燃料和废物的燃烧，现在随着含铅汽油的逐步淘汰而减少

(2) 粉煤灰。

化石燃料燃烧残留物。

(3) 石棉。

1) 以前是由于刹车片磨损等原因引起的问题

2) 现在逐步淘汰通用。

重要名词和术语解析

Asbestos: Asbestos is the name given to a group of fibrous silicate minerals, typically those of the serpentine group, approximate formula $Mg_3P(Si_2O_5)(OH)_4$.

石棉：石棉是一组纤维状硅酸盐矿物的名称，通常为蛇纹石族，近似式为 $Mg_3P(Si_2O_5)(OH)_4$。

2.2.6 Toxic Metals in the Atmosphere
大气中的有毒金属

知识框架

(1) Mercury.

1) On particles.

2) Hg(0) vapor, $Hg(CH_3)_2$.

(2) Lead.

1) One of 6 Priority Pollutants regulated by U.S. EPA.

2) Industrial sources.

3) Fossil fuel combustion.

4) Formerly as lead halides from leaded gasoline combustion.

(1) 汞。

1) 在颗粒上。

2) 汞蒸气，$Hg(CH_3)_2$。

(2) 铅。

1) EPA 受美国 EPA 管制的 6 种优先污染物之一。

2) 工业来源。

3) 化石燃料燃烧。

4) 以前是含铅汽油燃烧中的卤化铅。

(3) Beryllium.

1) High toxicity.

2) Lowest allowable limits of all elements.

3) Very limited uses

2.2.7 Radioactive Particles
放射性粒子

知识框架

(1) Some from cosmic rays acting on nuclei in atmosphere.

Example: ^{22}Na.

(2) Most troubling from natural sources is radon gas.

Decays to polonium which adheres to particles.

(3) Much particulate radioactivity from 1986 Chernobyl fire.

(4) Formerly many radioisotopes in the atmosphere and fallout from above-ground nuclear weapons testing.

重要名词和术语解析

Radon: A significant natural source of radionuclides in the atmosphere is **radon**, a noble gas product of radium decay.

2.2.8 Organic Particles in the Atmosphere
大气中的有机粒子

知识框架

(1) Hydrocarbons.

1) Long-chain alkanes cause few problems.

2) Aromatics more toxic.

（3）铍。

1）高毒性。

2）所有元素中允许极限最低。

3）非常有限的用途。

（1）一些来自宇宙射线作用于大气中的原子核。

例：^{22}Na。

（2）来自自然资源的最令人不安的是氡气。

衰变为附着在颗粒上的钋。

（3）1986年切尔诺贝利大火产生许多微粒放射性。

（4）以前大气中有许多放射性同位素以及地核武器试验的后果。

氡：氡是大气中放射性核素的重要自然来源，是镭衰变的稀有气体。

（1）碳氢化合物。

1）长链烷烃几乎没有问题。

2）芳烃毒性更大。

(2) Organooxygen particles.

1) Atmospheric oxidation products of hydrocarbons.

2) Aldehydes.

3) Ketones.

4) Carboxylic acids.

(3) Organonitrogen compounds such as acridine.

(4) Hydrocarbons and their derivatives are bound to carbonaceous particles from diesel engine emissions.

Collected by filters and burned off.

Polycyclic Aromatic Hydrocarbons and their Oxidation Products in Organic Particulate Matter.

2.2.9 Effects of Particles
粒子的影响

知识框架

(1) Particles are one of six Criteria Pollutants designated by the U.S. Environmental Protection Agency.

Others are sulfur dioxide, carbon monoxide, nitrogen dioxide, and lead.

(2) Most obvious effect is reduction and distortion of visibility.

Most pronounced in $0.1 \sim 1 \mu m$ range near wavelengths of visible light.

(3) Health effects from respirable particles.

Especially those less than $2.5 \mu m$, $PM_{2.5}$.

(4) Elevated levels of particles in the 1952 London air pollution episode.

4000 more deaths than usual over 5 days.

(2) 有机氧颗粒。

1) 碳氢化合物的大气氧化产物。

2) 醛。

3) 酮。

4) 羧酸。

(3) 有机氮化合物,如吖啶。

(4) 碳氢化合物及其衍生物与柴油机排放的碳质颗粒结合。

被过滤器收集并烧掉。

有机微粒中的多环芳烃及其氧化产物。

(1) 颗粒物是美国环境保护署指定的六种标准污染物之一。

其他是二氧化硫、一氧化碳、二氧化氮和铅。

(2) 最明显的影响是可见度的降低。

在可见光波长附近 $0.1 \sim 1 \mu m$ 最明显。

(3) 可吸入颗粒物对健康的影响。

特别是那些小于 $2.5 \mu m$ 的粒子,$PM_{2.5}$。

(4) 1952 年伦敦空气污染事件中颗粒物含量升高。

5 天之内的死亡人数比平常多 4000 人。

2.2.10 Water as Particulate Matter
水作为颗粒物

知识框架

(1) Water droplets are widespread in the atmosphere.
1) In clouds.
2) In fog.
(2) Water droplets in fog as carriers of pollutants.
1) Strong acid, especially H_2SO_4.
2) Corrosive salts, especially ammonium sulfate and nitrate.
(3) Most important effect is visibility reduction.
(4) Water droplets are important media for atmospheric chemical processes.
1) Oxidation of SO_2 to H_2SO_4.
2) Reactions involving HO·.
①Transferred from air to water.
②Produced chemically or photochemically in water.

(1) 水滴散布在大气中。
1) 在云里。
2) 在雾里。
(2) 雾中的水滴作为污染物的载体。
1) 强酸,尤其是硫酸。
2) 腐蚀性盐,尤其是硫酸铵和硝酸盐。
(3) 最重要的作用是降低能见度。
(4) 水滴是大气化学过程的重要介质。
1) SO_2 氧化为 H_2SO_4。
2) 涉及 HO· 的反应。
①从空气转移到水;
②在水中通过化学或光化学产生。

2.2.11 Control of Particle Emissions
颗粒物排放的控制

知识框架

(1) Particle control by sedimentation (larger particles).
(2) Inertial mechanisms that force particles onto a surface.
(3) Inertial mechanisms for particle removal. Dry centrifugal collectors (cyclones).

(1) 通过沉降控制颗粒(较大的颗粒)。
(2) 惯性机制,将颗粒压到表面上。
(3) 惯性机理用于去除颗粒。干式离心收集器(旋风分离器)。

(1) Filtration.

(2) Scrubbers.

1) Use liquids to wash particles from gas streams.

2) Ionizing wet scrubbers place a charge on particles prior to scrubbing.

(1) 过滤。

(2) 洗涤塔。

1) 用液体冲洗气流中的颗粒。

2) 在洗涤之前, 电离湿式洗涤器会在颗粒上带电。

重要名词和术语解析

Sedimentation: The simplest means of particulate matter removal is **sedimentation**, a phenomenon that occurs continuously in nature.

Inertial mechanism: Inertial mechanisms are effective for particle removal. These depend upon the fact that the radius of the path of a particle in a rapidly moving, curving air stream is larger than the path of the stream as a whole.

Dry centrifugal collector: When a gas stream is spun by vanes, a fan, or a tangential gas inlet, the particulate matter may be collected on a separator wall because the particles are forced outward by centrifugal force. Devices utilizing this mode of operation are called **dry centrifugal collectors** (cyclones).

Fabric filter: **Fabric filters**, as their name implies, consist of fabrics that allow the passage of gas but retain particulate matter.

Ionizing wet scrubber: **Ionizing wet scrubbers** place an electrical charge on particles upstream from a wet scrubber.

沉降: 去除颗粒物最简单的方法是**沉降**, 这是自然界中连续发生的现象。

惯性机制: **惯性机制**可有效去除颗粒。取决于以下事实: 在快速移动的弯曲空气流中, 粒子的路径半径比整个粒子流的路径大。

干式离心收集器: 当气流通过叶片、风扇或切向进气口旋转时, 颗粒物质可能会被收集在分离器壁上, 因为颗粒会由于离心力而被迫向外排出。利用这种操作模式的设备称为**干式离心收集器**(旋风分离器)。

织物过滤器: 顾名思义, **织物过滤器**由允许气体通过但截留颗粒物的织物组成。

电离湿式洗涤器: **电离湿式洗涤器**会在湿式洗涤器上游的颗粒上带电荷。

2.3 Gaseous Inorganic Pollutants
气态无机污染物

2.3.1 Inorganic Pollutant Gases
无机污染物气体

知识框架

(1) Aside from CO_2, most common from anthrosphere sources are:
1) CO.
2) SO_2.
3) NO.
4) NO_2.

(2) Others from pollutant sources:
1) NH_3.
2) N_2O.
3) N_2O_5.
4) H_2S.
5) Cl_2.
6) HCl.
7) HF.

(3) Large quantities from natural sources:
1) NH_3.
2) N_2O.
3) H_2S.

(1) 除了 CO_2 以外，最常见的来自人类活动圈的是。
1) CO;
2) SO_2;
3) NO;
4) NO_2。

(2) 来自其他污染源：
1) NH_3;
2) N_2O;
3) N_2O_5;
4) H_2S;
5) Cl_2;
6) HCl;
7) HF。

(3) 大量来自自然资源：
1) NH_3;
2) N_2O;
3) H_2S。

2.3.2 Production and Control of Carbon Monoxide
一氧化碳的产生与控制

知识框架

(1) About 0.1 parts per million by volume in atmosphere.
1) Residence time of 36~110 days.

(1) 一氧化碳在大气中的体积分数约 $0.1×10^{-7}$。
1) 存留时间 36~110 天。

2) About 2/3 CO is intermediate in oxidation of CH_4 by $HO\cdot$.

(2) Most common pollutant problem is localized in urban areas from automotive emissions.

Controlled by computerized engine control and exhaust catalysts.

2）大约 2/3 的 CO 是 $HO\cdot$ 氧化 CH_4 的中间物。

（2）最常见的污染问题是由于汽车排放导致的，局限于城市地区。

由计算机控制的发动机控制和排气催化剂控制。

2.3.3 Fate of Atmospheric CO
大气中一氧化碳的归宿

知识框架

(1) Oxidized by reaction with hydroxyl radical:

$$CO+HO\cdot \longrightarrow CO_2+H$$

(2) Production of hydroperoxyl radical:

$$O_2+H+M \longrightarrow HOO\cdot +M$$

(3) Additional reactions:

$$HOO\cdot +NO \longrightarrow HO\cdot +NO_2$$
$$HOO\cdot +HOO\cdot \longrightarrow H_2O_2+O_2$$
$$H_2O_2+h\nu \longrightarrow HO\cdot$$

(4) Soil microorganisms metabolize CO. Soil is a sink for CO.

（1）与羟基自由基反应被氧化：

（2）过氧羟基自由基的产生：

（3）其他反应：

（4）土壤微生物代谢 CO。土壤是 CO 的汇。

2.3.4 Sulfur dioxide Sources and the Sulfur Cycle
二氧化硫的来源和硫循环

知识框架

(1) Of the order of 100 million tons of sulfur enter the atmosphere from the anthrosphere each year.

Primarily as SO_2 from combustion of sulfur-containing fossil fuels.

（1）每年大约有 1 亿吨硫从人类活动圈进入大气。

主要是燃烧含硫化石燃料产生的 SO_2。

(2) Fluxes of sulfur from the geosphere and biosphere are large but uncertain.

1) H_2S and SO_2 from volcanic activity.

2) H_2S from action of anoxic bacteria on organosulfur compounds and SO_4^{2-}.

3) $(CH_3)_2S$ from marine microorganisms.

2.3.5 Sulfur Dioxide Reactions in the Atmosphere 大气中的二氧化硫反应

知识框架

(1) SO_2 is oxidized to sulfate SO_2.

1) H_2SO_4 in aerosol droplets.

2) Sulfate salts, predominantly NH_4HSO_4 and $(NH_4)_2SO_4$.

(2) Sulfuric acid and sulfates account for much of the atmospheric haze.

(3) Oxidation of SO_2 under various conditions.

1) Very rapid under oxidizing conditions that occur in presence of N oxides, hydrocarbons, and sunlight (photochemical smog conditions).

2) In solution inside water aerosol droplets.

3) On catalytic solid surfaces.

(4) Hydroxyl radicals react with SO_2 leading to oxidation.

(5) Other atmospheric oxidants of SO_2:

1) Ozone, O_3.

2) Hydrogen peroxide, H_2O_2.

(6) Metal ions catalyze SO_2 oxidation:

1) Fe(Ⅲ).

2) Mn(Ⅱ).

(2) 来自地圈和生物圈的硫通量很大，但不确定。

1) 火山活动中的 H_2S 和 SO_2。

2) 缺氧细菌对有机硫化合物和 SO_4^{2-} 的作用产生 H_2S。

3) 来自海洋微生物的 $(CH_3)_2S$。

(1) SO_2 被氧化成硫酸盐。

1) 气溶胶液滴中的 H_2SO_4。

2) 硫酸盐，主要是 NH_4HSO_4 和 $(NH_4)_2SO_4$。

(2) 硫酸和硫酸盐占大气雾霾的大部分。

(3) 在各种条件下氧化 SO_2。

1) 在存在氮氧化物、碳氢化合物和太阳光的氧化条件下（光化学烟雾条件），反应非常迅速。

2) 在水气溶胶滴中的溶液中。

3) 在催化固体表面上。

(4) 羟基与 SO_2 反应导致氧化。

(5) SO_2 的其他大气氧化剂：

1) 臭氧，O_3；

2) 过氧化氢，H_2O_2。

(6) 金属离子催化 SO_2 氧化：

1) Fe(Ⅲ)；

2) Mn(Ⅱ)。

Effects of Atmospheric Sulfur Dioxide

(1) Not particularly toxic.

People with some respiratory conditions (asthma) are sensitive to SO_2 exposure.

(2) Incidents of acute air pollution have been associated with high SO_2 levels.

1) 1930, Meuse River Valley, Belgium.

2) 1948, Donora, Pennsylvania.

3) 1952, London.

(3) Phytotoxic to some plants.

Leaf chlorosis (bleaching).

(4) Erosion of surfaces of stone, such as dolomite:

$$CaCO_3 \cdot MgCO_3 + 2SO_2 + O_2 + 9H_2O \longrightarrow CaSO_4 \cdot 2H_2O + MgSO_4 \cdot 7H_2O + 2CO_2$$

Sulfur Dioxide Removal

(1) Removed from coal combustion products.

(2) Fluidized bed combustion of coal in a bed of granular material that sequesters SO_2.

(3) SO_2 removal from stack gas.

(4) Dry processes using alkaline sorbents, such as $Ca(OH)_2$, are not very effective.

(5) Lime slurry scrubbing of stack gas is most used:

1) Wet throwaway process.

2) $Ca(OH)_2 + SO_2 \rightarrow CaSO_3 + H_2O$.

3) $CaSO_3$ may be oxidized to produced gypsum, $CaSO_4 \cdot 2H_2O$.

(6) Wet regenerative systems use alkaline absorbent, such as Na_2SO_3 solution, from which SO_2 is recovered.

大气二氧化硫的影响

(1) 不是特别有毒。

患有呼吸道疾病（哮喘）的人对 SO_2 暴露敏感。

(2) 急性空气污染事件与 SO_2 含量高有关。

1) 1930 年，比利时默兹河谷。

2) 1948 年，宾夕法尼亚州多诺拉。

3) 1952 年，伦敦。

(3) 对某些植物有植物毒性。

叶褪绿（漂白）。

(4) 石材表面的侵蚀，如白云石：

二氧化硫的去除

(1) 从煤燃烧产品中去除。

(2) 煤在隔离 SO_2 的颗粒状物料床中流化床燃烧。

(3) 从烟道气中去除 SO_2。

(4) 使用碱性吸附剂，如 $Ca(OH)_2$ 的干法效果不佳。

(5) 最常使用石灰浆洗涤烟气：

1) 湿式一次性处理；

2) $Ca(OH)_2 + SO_2 \rightarrow CaSO_3 + H_2O$；

3) $CaSO_3$ 可能被氧化成石膏，$CaSO_4 \cdot 2H_2O$。

(6) 湿式再生系统使用碱性吸收剂（如 Na_2SO_3 溶液）从中回收 SO_2。

1) Some SO_2 reduced to H_2S.

2) Claus reaction to produce S: $2H_2S+SO_2 \rightarrow 3S+2H_2O$ (Example of a green process).

Oxy-Fuel Combustion for Sulfur Dioxide and Carbon Dioxide Recovery

Pure oxygen to support combustion (oxyfuel combustion) produces an exhaust that is highly enriched in SO_2 and CO_2, greatly facilitating their removal from the exhaust.

(1) Requires exhaust gas recirculation to prevent overheating.

(2) Saves energy by not heating large volumes of N_2 in air.

(3) Energy conserved by condensing water vapor in exhaust.

(4) Much smaller volumes of exhaust gas require treatment.

(5) No NO from N_2 reaction with O_2, but some from N in coal.

2.3.6 Nitrogen Oxides in the Atmosphere 大气中的氮氧化物

知识框架

(1) N_2O emitted in relatively large quantities from natural and some pollutant sources.

(2) NO from combustion, especially internal combustion engine, lightning.

(3) NO_2, largely secondary from NO oxidation.

(4) NO_3 radical, secondary pollutant.

NO and NO_2 are collectively termed NO_x.

1) 一些 SO_2 还原为 H_2S。

2) 克劳斯反应生成 S: $2H_2S+SO_2 \rightarrow 3S+2H_2O$（绿色过程示例）。

氧燃料燃烧用于二氧化硫和二氧化碳的回收

支持燃烧的纯氧（氧气燃料燃烧）产生的废气中富含 SO_2 和 CO_2，极大地促进了它们从废气中的去除。

(1) 需要废气循环以防止过热。

(2) 通过不加热空气中的大量 N_2 来节省能源。

(3) 通过冷凝废气中的水蒸气节省能量。

(4) 更少的废气需要处理。

(5) N_2 与 O_2 反应中不产生 NO，但煤中的 N 会产生一些。

(1) 从自然和某些污染物源中排放出相对大量的 N_2O。

(2) 燃烧（尤其是内燃机）、闪电产生 NO。

(3) NO_2 主要来自 NO 氧化。

(4) NO_3 自由基，二次污染物。

NO 和 NO_2 统称为 NO_x。

One of the six criteria pollutants regulated by U. S. Environmental Protection Agency.

Some tens to over 100 million tons of NO_x emitted to the atmosphere annually.

Harmful Effects of Nitrogen Oxides

(1) NO relatively insignificant.

(2) NO_2 harmful to health.

Fatal bronchiolitis fibrosa obliterans.

(3) NO_2 leads to formation of corrosive and harmful nitric acid and nitrates.

(4) Most significant reaction of atmospheric NO_2 is the following producing O atoms that participate in important atmospheric chain reactions:

1) $NO_2 + h\nu \rightarrow NO + O$.

2) Leads to harmful photochemical smog formation.

3) NO_x often limiting reactant in production of photochemical smog.

Control of Nitrogen Oxides

(1) Prevent production by control of combustion conditions.

(2) Removal from exhaust gas after combustion.

Prevention of production.

(1) Low excess air firing to limit O in $N_2 + O_2 \rightarrow 2NO$.

(2) Low combustion temperatures prevent NO formation.

Removal from exhaust gas.

(1) Sorption by base relatively ineffective due to low acidity of N oxides.

美国环境保护署规定的六种标准污染物之一。

每年排放到大气中的 NO_x 有数十吨至一亿吨。

氮氧化物的有害影响

(1) NO 相对微不足道。

(2) NO_2 对健康有害。

致命性纤维化闭塞性细支气管炎。

(3) NO_2 导致腐蚀性和有害硝酸和硝酸盐的形成。

(4) 大气中 NO_2 最重要的反应是以下产生的 O 原子，它们参与重要的大气链反应：

1) $NO_2 + h\nu \rightarrow NO + O$；

2) 导致有害的光化学烟雾形成；

3) NO_x 经常限制光化学烟雾产生中的反应物。

氮氧化物的控制

(1) 通过控制燃烧条件防止产生。

(2) 从燃烧后的废气中去除。

预防产生。

(1) 低过量空气燃烧以限制 $N_2 + O_2 \rightarrow 2NO$ 中的 O。

(2) 燃烧温度低，防止形成 NO。

从废气中去除。

(1) 由于 N 氧化物的酸度较低，因此对碱的吸附相对无效。

(2) Most commonly by catalytic reduction with reducing agents.

(3) Automotive exhaust catalysts reduce NO_x with slight excess CO and hydrocarbons.

Biofilters consisting of microorganisms on fixed or fluidized supports are an experimental approach to NO_x control.

2.3.7 Acid Rain
酸雨

知识框架

Acidic precipitation from strong acids

(1) HNO_3 secondary pollutant from NO_x.

(2) H_2SO_4 secondary pollutant from SO_2.

(3) HCl typically from combustion of organohalide polymers.

2.3.8 Ammonia in the Atmosphere
大气中的氨

知识框架

(1) Numerous sources of atmospheric NH_3.
1) Soil microorganisms.
2) Decay of animal wastes.
3) Ammonia fertilizer.
4) Sewage treatment.
5) NH_3 synthesis NH_3.
6) Leakage of ammonia-based refrigeration systems.
7) Livestock and feedlot operations largest

(2) 最常见的是通过还原剂催化还原。

(3) 汽车尾气催化剂可通过少量过量的 CO 和碳氢化合物还原 NO_x。

由固定或流化支持物上的微生物组成的生物滤池是控制 NO_x 的实验方法。

强酸的酸沉降

(1) 产生的 HNO_3 二次污染物 NO_x。

(2) SO_2 产生的 H_2SO_4 二次污染物。

(3) HCl 通常来自有机卤化物聚合物的燃烧。

(1) 大气中 NH_3 的大量来源。
1) 土壤微生物；
2) 动物粪便的腐烂；
3) 氨肥；
4) 污水处理；
5) 合成；
6) 氨制冷系统泄漏；
7) 牲畜和饲养场运营是

U. S. source (perhaps as much as 1 billion kg per year).

(2) Ammonia is the major base in the atmosphere.

(3) Ammonia removed by its water solubility and basicity.

1) With rain as $NH_3(aq)$.

2) By reaction with H_2SO_4, HNO_3, HCl.

Effects of atmospheric ammonia

(1) Ammonium sulfate major atmospheric aerosol—visibility.

(2) Ammonium salts cause corrosion.

(3) Damage to foliage.

2.3.9 Fluorine, Chlorine and their Gaseous Compounds 氟、氯及其气态化合物

知识框架

(1) F_2 and HF are rare air pollutants.

1) Extremely toxic.

2) React even with glass.

(2) Plants susceptible to fluorides.

(3) Sulfur hexafluoride, SF_6, is ultrastable.

Very strong greenhouse gas.

(4) Chlorine and Hydrogen Chloride. Occasional spills of Cl_2 still cause fatalities.

(5) Hydrogen chloride, HCl, rapidly forms hydrochloric acid droplets in the atmosphere.

Major acid rain constituent.

美国最大的来源（每年可能多达 10 亿千克）。

（2）氨是大气中的主要碱。

（3）通过其水溶性和碱性去除氨。

1）以 NH_3 的水溶液存在于雨水中；

2）通过与 H_2SO_4、HNO_3、HCl 反应。

大气氨的影响

（1）硫酸铵主要大气气溶胶——能见度。

（2）铵盐会引起腐蚀。

（3）破坏树叶。

（1）F_2 和 HF 是稀有的空气污染物。

1）剧毒；

2）甚至与玻璃反应。

（2）植物易受氟化物影响。

（3）六氟化硫 SF_6 超级稳定。

非常强的温室气体。

（4）氯和氯化氢。

偶尔发生的 Cl_2 泄漏仍会导致死亡。

（5）氯化氢 HCl 在大气中迅速形成盐酸滴。

主要酸雨成分。

2.3.10 Reduced Sulfur Gases
还原的硫气体

知识框架

(1) Major reduced sulfur gases in the atmosphere.
1) Hydrogen sulfide, H_2S.
2) Carbonyl sulfide, OCS.
3) Carbon disulfide, CS_2.
4) Dimethyl sulfide, $S(CH_3)_2$.

(2) Hydrogen sulfide is most damaging reduced sulfur gas in the atmosphere.
1) About as toxic as hydrogen cyanide.
2) Phytotoxic.
3) Damage to materials.

Formerly lead paint pigments.

(1) 大气中主要还原的硫气体有：
1) 硫化氢，H_2S；
2) 羰基硫化物，OCS；
3) 二硫化碳，CS_2；
4) 二甲基硫，$S(CH_3)_2$。

(2) 硫化氢是大气中最具破坏性的还原硫气体。
1) 毒性与氰化氢一样。
2) 植物毒性。
3) 材料损坏。

以前的含铅颜料。

2.4 Organic Air Pollutants
有机大气污染物

2.4.1 Organic Compounds in the Atmosphere
大气中的有机化合物

知识框架

(1) Direct effects.
Example: Cancer from vinyl chloride.
(2) Secondary pollutants.
Especially photochemical smog.
(3) Persistent Organic Pollutants are poorly biodegradable organic compounds recognized as especially troublesome.
1) Last in the Earth System for many years.

(1) 直接效果。
例：氯乙烯引起的癌症。
(2) 二次污染物。
特别是光化学烟雾。
(3) 持久性有机污染物是难降解的有机化合物，被认为特别麻烦。
1) 在地球系统中持续多年。

2) Become widely distributed throughout the Earth System.

3) Accumulate in fatty tissue of organisms, including humans, in especially higher concentrations at higher levels in the food chain.

4) Are toxic to humans and wildlife.

Examples: Chlordane, hexachlorobenzene, PCBs.

(4) Loss of Organic Substances from the Atmosphere.

1) Precipitation (rainwater).

2) Dry deposition.

3) Photochemical reactions.

4) Incorporation into particles.

5) Tend to undergo photochemical reactions leading to solids that are purged from the atmosphere.

6) Uptake by plants, especially trees.

Absorbed by lipophilic layer on leaves and needles of trees.

2) 广泛分布于整个地球系统。

3) 在包括人类在内的生物的脂肪组织中积累, 尤其是食物链等级越高的生物体内浓度越高。

4) 对人类和野生生物有毒。

例如: 氯丹, 六氯苯, 多氯联苯。

(4) 大气中有机物的流失。

1) 降水 (雨水)。

2) 干沉降。

3) 光化学反应。

4) 掺入颗粒。

5) 倾向于发生光化学反应, 导致从大气中清除固体。

6) 植物, 特别是树木的吸收。

被树叶和树针上的亲脂层吸收。

2.4.2 Biogenic Organic Compounds in the Atmosphere 大气中的生物有机化合物

知识框架

(1) Natural sources most abundant sources of atmospheric organics.

1) Methane from bacteria and geosphere is the most abundant organic in the atmosphere.

2) Anoxic bacteria: $2\{CH_2O\} \rightarrow CH_4 + CO_2$.

3) Flatulent emissions from livestock.

(2) Terpenes from vegetation, primarily

(1) 天然来源是大气有机物的最丰富来源。

1) 来自细菌和地圈的甲烷是大气中最丰富的有机物。

2) 缺氧细菌: $2\{CH_2O\} \rightarrow CH_4 + CO_2$。

3) 牲畜的肠胃气态排放。

(2) 来自植被的萜烯, 主

pine and citrus trees, are second to methane as organics in the atmosphere.

1) Generally very reactive (in photochemical smog formation).

2) Form much of the small particulate matter in atmosphere.

(3) Esters in the Atmosphere.

Many kinds of esters, largely from plant sources.

(4) Removal of Atmospheric Organic Compounds by Plants.

1) Repositories of persistent organic pollutants.

2) Leaves and needles covered by epicular organophilic wax that accumulates atmospheric organics.

① Especially in evergreen boreal coniferous forests in the northern temperate zone.

② Heavy forestation and large leaf surface per unit forest area.

要是松树和柑橘树，是大气中仅次于甲烷的有机物。

1）通常非常活跃（在光化学烟雾形成中）。

2）在大气中形成许多小颗粒物。

(3）大气中的酯。

多种酯，主要来自植物。

(4）植物去除大气有机化合物。

1）持久性有机污染物库。

2）叶和针被表层亲有机蜡覆盖，可积累大气有机物。

①特别是在北部温带地区常绿的北方针叶林中。

②茂密的造林和单位森林面积的大叶面。

重要名词和术语解析

Biogenic hydrocarbon: Atmospheric hydrocarbons produced by living sources are called **biogenic hydrocarbons**.

Terpene: Most of the hydrocarbons emitted by plants are **terpenes**, which constitute a large class of organic compounds found in essential oils.

生物碳氢化合物：生物来源产生的大气碳氢化合物称为**生物碳氢化合物**。

萜烯：植物排放的大多数碳氢化合物是**萜烯**，它们构成了香精油中的一类有机化合物。

2.4.3 Pollutant Hydrocarbons
碳氢化合物污染物

知识框架

(1) Alkanes, such as 2,2,3-Trimethylbutane.

（1）烷烃，比如2,2,3-三甲基丁烷。

(2) Alkanes are relatively very stable.
(3) Alkanes undergo abstraction reactions.
1) React with O from NO_2 photodissociation.

2) React with HO·.

3) Additional reactions that produce solid aerosols and soluble substances that are removed from the atmosphere.

（2）烷烃相对非常稳定。
（3）烷烃经历提取反应。
1）与 NO_2 光离解中的 O 反应。

2）与 HO· 反应。

3）产生固体气溶胶和可溶性物质的其他反应，从大气中清除。

重要名词和术语解析

Alkene：olefins, compounds with double bonds between adjacent carbon atoms, such as ethylene.

Alkyne：compounds with triple bonds, such as acetylene.

Aryl compound：Aryl (**aromatic**) **compounds**, such as naphthalene.

Alkyl radical：In the atmosphere, alkanes (general formula C_xH_{2x+2}) are attacked primarily by hydroxyl radical, HO·, resulting in the loss of a hydrogen atom and formation of an **alkyl radical**. These radicals may act as oxidants, losing oxygen (usually to NO forming NO_2) to produce **alkoxyl radicals**.

Aryl hydrocarbon：Aryl (**aromatic**) **hydrocarbons** may be divided into the two major classes of those that have only one benzene ring and those with multiple rings.

Delocalized：The electron is not confined to one atom; therefore, it is **delocalized** and may be represented in the aryl radical structure by a half-circle with a dot in the middle.

Formaldehyde：The simplest and most widely

烯烃：烯烃，在相邻碳原子之间具有双键的化合物，如乙烯。

炔烃：具有三键的化合物，如乙炔。

芳基化合物：芳基（芳族）化合物，如萘。

烷基自由基：在大气中，烷烃（通式 C_xH_{2x+2}）主要受到羟基 HO· 的攻击，导致氢原子的损失和**烷基自由基**的形成。这些自由基可作为氧化剂，失去氧气（通常会生成 NO 生成 NO_2），从而生成**烷氧基自由基**。

芳烃：芳烃（芳族）可分为两大类，即仅具有一个苯环的烃和具有多个环的烃。

离域：电子不局限于一个原子；因此它是**离域**的，可以在芳基结构中以半圆表示，中间带点。

甲醛：最简单，生产最广泛的

produced of the carbonyl compounds is the lowest aldehyde.

Chromophore: a molecular group that readily absorbs light.

Organohalide: **Organohalides** consist of halogen-substituted hydrocarbon molecules, each of which contains at least one atom of F, Cl, Br, or I. They may be saturated (**alkyl halides**), unsaturated (**alkenyl halides**), or aryl (**aryl halides**).

Chloromethane: Volatile **chloromethane** (methyl chloride) is consumed in the manufacture of silicones.

Dichloromethane: **Dichloromethane** is a volatile liquid with excellent solvent properties for nonpolar organic solutes. It has been used as a solvent for the decaffeination of coffee, in paint strippers, as a blowing agent in urethane polymer manufacture, and to depress vapor pressure in aerosol formulations.

Dichlorodifluoromethane: **Dichlorodifluoromethane** is one of the chlorofluorocarbon compounds once widely manufactured as a refrigerant and involved in stratospheric ozone depletion. One of the more common industrial chlorinated solvents is 1, 1, 1-trichloroethane.

Alkenyl: Viewed as halogen-substituted derivatives of alkenes, the **alkenyl** or **olefinic organohalides** contain at least one halogen atom and at least one carbon-carbon double bond.

羰基化合物是最低级的醛。

发色团：易于吸收光的分子基团。

有机卤化物：**有机卤化物**由卤素取代的烃分子组成，每个烃原子包含至少一个 F、Cl、Br 或 I 原子。它们可以是饱和的（**烷基卤化物**），不饱和的（**烯基卤化物**）或芳基（**芳基卤化物**）。

氯甲烷：制造有机硅时会消耗挥发性**氯甲烷**（甲基氯化物）。

二氯甲烷：**二氯甲烷**是一种挥发性液体，对非极性有机溶质具有优异的溶剂性能。它已被用作咖啡脱咖啡因的溶剂、脱漆剂、氨基甲酸酯聚合物生产中的发泡剂以及降低气雾剂的蒸气压。

二氯二氟甲烷：**二氯二氟甲烷**是一种曾经广泛用作制冷剂并涉及平流层臭氧消耗的氯氟烃化合物之一。一种较常见的工业氯化溶剂是 1,1,1－三氯乙烷。

烯基：被视为烯烃的卤素取代衍生物，**烯基**或**烯属有机卤化物**包含至少一个卤素原子和至少一个碳—碳双键。

Vinyl chloride: Vinyl chloride is consumed in large quantities as a raw material to manufacture pipe, hose, wrapping, and other products fabricated from polyvinyl chloride plastic.

Trichloroethylene: Trichloroethylene is a clear, colorless, nonflammable, volatile liquid.

Allyl chloride: Allyl chloride is an intermediate in the manufacture of allyl alcohol and other allyl compounds, including pharmaceuticals, insecticides, and thermosetting varnish and plastic resins.

Chlorofluorocarbon: Chlorofluorocarbons (**CFCs**), such as dichlorodifluoromethane, commonly called Freons, are volatile 1- and 2-carbon compounds that contain Cl and F bonded to carbon.

Halon: Halons are related compounds that contain bromine and are used in fire extinguisher systems.

Perfluorocarbon: Perfluorocarbons are completely fluorinated organic compounds, the simplest examples of which are carbon tetrafluoride (CF_4) and hexafluoroethane (C_2F_6).

Amine: Amines consist of compounds in which one or more of the hydrogen atoms in NH_3 has been replaced by a hydrocarbon moiety.

Heterocyclic nitrogen compound: A large number of **heterocyclic nitrogen compounds** have been reported in tobacco smoke, and it is inferred that many of these compounds can enter the atmosphere from burning vegetation.

氯乙烯：氯乙烯作为原材料大量消耗，用于制造管道、软管、包装材料和其他由聚氯乙烯塑料制成的产品。

三氯乙烯：三氯乙烯是一种透明、无色、不可燃的挥发性液体。

烯丙基氯：烯丙基氯是制造烯丙醇和其他烯丙基化合物（包括药品、杀虫剂、热固性清漆和塑料树脂）的中间体。

氯氟烃：氯氟烃（**CFCs**），例如二氯二氟甲烷，通常称为氟利昂，是挥发性的1和2碳化合物，含有与碳键合的Cl和F。

卤代烷：卤代烷是含有溴的相关化合物，用于灭火器系统。

全氟化碳：全氟化碳是完全氟化的有机化合物，最简单的例子是四氟化碳（CF_4）和六氟乙烷（C_2F_6）。

胺：胺由NH_3中一个或多个氢原子被烃部分取代的化合物组成。

杂环氮化合物：烟草烟雾中已报告了许多**杂环氮化合物**，据推测，这些化合物中有许多可以从燃烧的植被进入大气。

Nitrosamine: Nitrosamines contain the N—N=O group.

亚硝胺：亚硝胺含有 N—N=O 基团。

2.5 Photochemical Smog
光化学烟雾

2.5.1 Introduction
简介

知识框架

(1) Originally, "smog" referred to reducing, sulfurous smog.

Generated from coal smoke and fog in London.

(2) Oxidizing photochemical smog characterized by:

1) Eye irritation.

2) Low visibility at low humidity.

3) Presence of oxidants including O_3.

(3) Photochemical smog forms in the troposphere's planetary boundary layer.

1) Extends up to about 1km.

2) Region of maximum interaction between tropospheric air and Earth's surface.

3) Location of temperature inversions in which photochemical smog forms.

2.5.2 Smog-Forming Emissions
烟雾形成排放

知识框架

(1) Automobile a prime source of smog forming emissions.

(2) Exhaust hydrocarbons, especially

(1) 最初，"烟雾"是指还原性的含硫烟雾。

来自伦敦的煤烟和雾气。

(2) 氧化光化学烟雾的特征是：

1) 刺激眼睛；

2) 低湿度下能见度低；

3) 存在氧化剂，包括臭氧。

(3) 在对流层的行星边界层中形成光化学烟雾。

1) 延伸约 1km；

2) 对流层空气与地球表面之间最大相互作用的区域；

3) 形成光化学烟雾的逆温的位置。

(1) 汽车是形成烟雾的主要来源。

(2) 排放的碳氢化合物，

unsaturated ones, are especially reactive in smog formation.

(3) Automobile also source of NO required for smog.

(4) Control of operational parameters of the four-cycle automobile engine important in smog control.

(5) Engine control to limit smog-forming emissions.

(6) Computerized control of timing, air/fuel ratio, other parameters limit emissions of NO, hydrocarbons (HC), and CO.

(7) Catalytic converters oxidize HC and CO and reduce NO.

Mixture cycles rapidly between slightly rich and slightly lean.

Trends in allowable automobile emissions (g/mile)❶

(1) Before controls:
　　HC, 10.6; CO, 84.0; NO_x, 4.1

(2) 1970:
　　HC, 4.1; CO, 34.0; NO_x, —

(3) 2008:
　　HC, 0.41; CO, 3.4; NO_x, 0.4

(4) After 2008: Continued reductions in emissions, diesel emissions regulated.

Polluting Green Plants

Plants are high contributors to reactive atmospheric HCs.

(1) Highly reactive terpenes such as α-pinene.

特别是不饱和碳氢化合物，在烟雾形成过程中尤其具有反应性。

(3) 汽车也是烟雾所需要的 NO 的来源。

(4) 控制烟雾控制中重要的四冲程汽车发动机的运行参数。

(5) 发动机控制以限制形成烟雾的排放。

(6) 计算机控制定时、空燃比，其他限制 NO、碳氢化合物（HC）、CO 的排放的参数。

(7) 催化转化器氧化 HC 和 CO 并减少 NO。

混合物在稍富和稍稀之间快速循环。

汽车允许排放量的趋势（克/英里）❶

(1) 控制前：

(2) 1970：

(3) 2008：

(4) 2008 年之后：继续减少排放，规定了柴油排放。

污染性的绿色植物

植物是反应性大气 HC 的重要贡献者。

(1) 高反应性萜烯，比如 α-蒎烯。

---

❶ 1mile（英里）= 1.609km。

（2）Most abundant is isoprene.

（3）Oxidized to carbonyls and other products.

（4）Isoprene nitrates from reactions with HO·, NO_x, NO_3 radical.

（2）最丰富的是异戊二烯。

（3）氧化成羰基化合物和其他产物。

（4）与HO·、NO_x、NO_3自由基反应产生的硝酸异戊二烯。

重要名词和术语解析

Intake：Air is drawn into the cylinder through the open intake valve. Gasoline is either injected with the intake air or injected separately into the cylinder.

Compression：The combustible mixture is compressed at a ratio of about 7∶1. Higher compression ratios favor thermal efficiency and complete combustion of hydrocarbons. However, higher temperatures, premature combustion ("pinging"), and high production of nitrogen oxides also result from higher combustion ratios.

Ignition and power stroke：As the fuel-air mixture normally produced by injecting fuel into the cylinder is ignited by the spark plug near top-dead-center, a temperature of about 2500℃ is reached very rapidly at pressures up to 40atm❶. As the gas volume increases with downward movement of the piston, the temperature decreases in a few milliseconds. This rapid cooling "freezes" nitric oxide in the form of NO without allowing it time to dissociate to N_2 and O_2, which are thermodynamically favored at the normal temperatures and pressures of the atmosphere.

进气：空气通过打开的进气门吸入气缸。汽油被注入进气或被单独注入气缸。

压缩：可燃混合物以大约7∶1的比率压缩。较高的压缩比有利于热效率和碳氢化合物的完全燃烧。但是，较高的燃烧比也会导致较高的温度，过早的燃烧（"砰声"）和大量产生氮氧化物。

点火和动力冲程：通过向气缸内注入燃料而产生的燃料-空气混合物，在上止点附近被火花塞点燃，在压力高达40atm❶时，温度很快达到约2500℃。气体量随活塞的向下移动而增加，温度在几毫秒内下降。这种快速冷却以NO形式"冻结"一氧化氮，而没有使其时间解离为N_2和O_2，这在常温和大气压下在热力学上是有利的。

---

❶ 1atm=101.325kPa。

Exhaust: Exhaust gases consisting largely of N_2 and CO_2, with traces of CO, NO, hydrocarbons, and O_2, are pushed out through the open exhaust valve, thus completing the cycle.

排气：通过打开的排气阀将主要由 N_2 和 CO_2 组成的排气以及痕量的 CO、NO、碳氢化合物和 O_2 排出，从而完成循环。

2.5.3 Smog-Forming Reactions of Organic Compounds in the Atmosphere
大气中有机化合物的烟雾形成反应

知识框架

Hydrocarbons undergo photochemical oxidation in the atmosphere to produce

（1）CO_2.

（2）Organic solids.

（3）Water-soluble aldehydes.

（4）Inorganic byproducts including O_3 and HNO_3.

碳氢化合物在大气中由光化学氧化反应产生

（1）CO_2。

（2）有机固体。

（3）水溶性醛。

（4）无机副产物，包括 O_3 和 HNO_3。

Reactions of methane to illustrate major kinds of smog-forming reactions

（1）$CH_4 + O$（from NO_2 dissociation）$\rightarrow H_3C\cdot + HO\cdot$.

An abstraction reaction involving the removal of an atom, usually H, by a reactive species such as O or HO·.

（2）Rapid reaction of hydroxyl radical：

$$CH_4 + HO\cdot \longrightarrow H_3C\cdot + H_2O$$

（3）$H_3C\cdot + O_2 + M \rightarrow H_3COO\cdot + M$

（4）Regeneration of NO_2, which can undergo further photodissociation：

$$H_3COO\cdot + NO \longrightarrow H_3CO\cdot + NO_2$$

（5）Production of hydroperoxyl radical：

$$H_3CO\cdot + O_2 \longrightarrow CH_2O + HOO\cdot$$

甲烷反应可以说明主要的烟雾形成反应

（1）$CH_4 + O$（来自 NO_2 分解）$\rightarrow H_3C\cdot + HO\cdot$

一种提取反应，通过反应性物质（如 O 或 HO·）除去一个通常为 H 的原子。

（2）羟基自由基的快速反应：

（3）$H_3C\cdot + O_2 + M \rightarrow H_3COO\cdot + M$。

（4）NO_2 的再生，可能会发生进一步的光解离：

（5）产生过氧羟基自由基：

(6) HO· and HOO· are odd hydrogen radicals that are ubiquitous intermediates in atmospheric chain reactions.

(7) CH_2O is photochemically active formaldehyde.

Addition Reactions of Unsaturated Compounds

(1) Addition of HO· across double bond:

$$HO\cdot + H-\underset{H}{\underset{|}{\overset{H}{\overset{|}{C}}}}-\underset{H}{\overset{H}{\overset{|}{C}}}=\underset{H}{\overset{H}{C}} \longrightarrow H-\underset{H}{\underset{|}{\overset{H}{\overset{|}{C}}}}-\underset{H}{\underset{|}{\overset{H}{\overset{|}{C}}}}-\underset{H}{\overset{H}{\overset{|}{C}}}-OH$$

(2) Addition reactions with ozone:

$$H-\underset{H}{\underset{|}{\overset{H}{\overset{|}{C}}}}-\underset{}{\overset{H}{\overset{|}{C}}}=\underset{H}{\overset{H}{C}} + O_3 \longrightarrow H-\underset{H}{\underset{|}{\overset{H}{\overset{|}{C}}}}-\overset{O-O}{\underset{O}{\overset{|\quad|}{C-C}}}\underset{H}{\overset{H}{}}$$

Primary photochemical reactions of organics, especially aldehydes:

$$H-\underset{H}{\underset{|}{\overset{H}{\overset{|}{C}}}}-\overset{O}{\overset{\|}{C}}-H + h\nu \longrightarrow H-\underset{H}{\overset{H}{\overset{|}{C}}}\cdot + H\dot{C}O$$

Reactions of Organic Free Radicals

Example: Generation of HO· from organic peroxyl radicals.

$$H-\underset{H}{\underset{|}{\overset{H}{\overset{|}{C}}}}-\overset{O\cdot}{\overset{|}{\underset{|}{\overset{O}{\overset{|}{C}}}}}-\underset{H}{\overset{H}{\overset{|}{C}}}-H \longrightarrow H-\underset{H}{\underset{|}{\overset{H}{\overset{|}{C}}}}-\overset{O}{\overset{\|}{C}}-\underset{H}{\overset{H}{\overset{|}{C}}}-H + HO\cdot$$

(1) Chain reactions with many steps.

(2) Hydroxyl radical key species in sustaining chain reactions.

(3) Chain branching.

(4) Chain termination.

(5) Two radicals react:

$$HO\cdot + HO\cdot \longrightarrow H_2O_2$$

（6）Radical adding to NO$_x$ (stable free radical):

$$HO\cdot + NO_2 + M \longrightarrow HNO_3 + M$$

（7）Radical adding to solid surface.

（6）自由基加成至 NO$_x$ (稳定的自由基):

（7）自由基加成至固体表面。

重要名词和术语解析

Abstraction reaction: abstraction reactions involving the removal of an atom, usually hydrogen, by reaction with an active species.

提取反应: 通过与活性物质反应除去一个原子，通常是氢。

2.5.4 Overview of Photochemical Smog Formation 光化学烟雾形成概述

知识框架

（1）Smog-forming conditions:
1) Hydrocarbon pollution.
2) NO pollution.
3) Intense sunlight.
4) Stagnant air.

（2）Photochemical smog evidenced by:
1) Gross photochemical oxidant that oxidizes I^- to I_3^-.
2) Main photochemical oxidant is ozone, O_3.
3) Other oxidants include:
① H_2O_2.
② Peroxides (ROOR′).
③ Organic hydroperoxides.
④ Peroxyacyl nitrates.

（1）烟雾形成条件:
1) 碳氢化合物污染;
2) NO 污染;
3) 强烈的阳光;
4) 停滞的空气;

（2）光化学烟雾出现的证明:
1) 将 I^- 氧化为 I_3^- 的总光化学氧化剂。
2) 主要光化学氧化剂是臭氧, O_3。
3) 其他氧化剂包括:
① H_2O_2;
② 过氧化物 (ROOR′);
③ 有机氢过氧化物 (ROOH);
④ 过氧酰基硝酸酯。

2.5.5 Mechanisms of Smog Formation 烟雾形成机理

知识框架

（1）Major kinds of reactions for smog formation.

（1）烟雾形成的主要反应类型。

1) Primary photochemical reaction producing oxygen atoms:

$$NO_2 + h\nu \longrightarrow NO + O$$

2) Reactions involving oxygen species:

$$O_2 + O + M \longrightarrow O_3 + M$$
$$O_3 + NO \longrightarrow O_2 + NO_2$$

3) Production of free radicals from hydrocarbons:

$$RH + O \longrightarrow R\cdot + HO\cdot$$
$$RH + O_3 \longrightarrow R\cdot + \text{other products}$$

4) Chain propagation, branching, and termination by a variety of reactions such as:

$$NO + ROO\cdot \longrightarrow NO_2 + \text{and/or other products}$$
$$NO_2 + R\cdot \longrightarrow \text{products (for example, PAN)}$$

(common chain-terminating reaction, NO_2 is a stable free radical species)

(2) Hydroxyl radical, HO·, is a very important species in propagating chains and generating products in photochemical smog.

1) $HO\cdot + NO_2 \rightarrow HNO_3$.

2) Oxidation of CO by hydroxyl radical: $HO\cdot + CO + O_2 \rightarrow CO_2 + HOO\cdot$.

Responsible for removal of CO and production of HOO·.

3) HOO· important in oxidation of NO to photochemically active NO_2: $HOO\cdot + NO \rightarrow NO_2 + HO\cdot$.

(3) Abstraction of H from alkanes leading to smog formation:

1) $RH + O + O_2 \longrightarrow ROO\cdot + HO\cdot$.

2) $RH + HO\cdot + O_2 \rightarrow ROO\cdot + H_2O$.

Addition reactions of HO· across double bonds in alkenes are very rapid:

1) 初级光化学反应产生氧原子:

$$NO_2 + h\nu \longrightarrow NO + O$$

2) 涉及氧的反应:

$$O_2 + O + M \longrightarrow O_3 + M$$
$$O_3 + NO \longrightarrow O_2 + NO_2$$

3) 由烃类 RH 产生自由基:

$$RH + O \longrightarrow R\cdot + HO\cdot$$
$$RH + O_3 \longrightarrow R\cdot + \text{其他产物}$$

4) 通过各种反应进行链的增长、分支和终止,例如:

$$NO + ROO\cdot \longrightarrow NO_2 + \text{和/或其他产物}$$
$$NO_2 + R\cdot \longrightarrow \text{产物(例如, PAN)}$$

(常见的链终止反应, NO_2 是稳定的自由基物质)

(2) 羟基自由基 HO· 是光化学烟雾中链增长和产生产物的重要物种。

1) $HO\cdot + NO_2 \rightarrow HNO_3$。

2) 羟基自由基氧化 CO: $HO\cdot + CO + O_2 \longrightarrow CO_2 + HOO\cdot$。

负责去除 CO 和产生 HOO·。

3) HOO· 在将 NO 氧化为光化学活性 NO_2 中很重要: $HOO\cdot + NO \rightarrow NO_2 + HO\cdot$。

(3) 从烷烃中提取 H 导致烟雾形成:

1) $RH + O + O_2 \rightarrow ROO\cdot + HO\cdot$。

2) $RH + HO\cdot + O_2 \rightarrow ROO\cdot + H_2O$。

HO· 跨烯烃双键的加成反应非常快:

$$\begin{CD}
\underset{R}{\overset{R}{>}}C=C\underset{R}{\overset{R}{<}} +HO\cdot @>\text{Very rapid 非常快速}>> HO-\underset{\underset{R}{|}}{\overset{\overset{R}{|}}{C}}-\underset{\underset{R}{|}}{\overset{\overset{R}{|}}{C}}\cdot @>>> \text{Oxidation products 氧化产物}
\end{CD}$$

Radical adduct 双自由基

$$\underset{R}{\overset{R}{>}}C=C\underset{R}{\overset{R}{<}} +O \longrightarrow R-\underset{\underset{R}{|}}{\overset{\overset{O\cdot}{|}}{C}}-\underset{\underset{R}{|}}{\overset{\overset{R}{|}}{C}}\cdot \longrightarrow \text{Oxidation products 氧化产物}$$

Radical adduct 双自由基

$$\underset{R}{\overset{R}{>}}C=C\underset{R}{\overset{R}{<}} +O_3 \longrightarrow R-\overset{O-O}{\overset{|\quad|}{\underset{\underset{R}{|}}{C}-\underset{\underset{R}{|}}{C}}}-R \longrightarrow \text{Oxidation products 氧化产物}$$

3) Because of addition reactions, alkenes are very reactive in photochemical smog formation.

3) 由于加成反应，烯烃在光化学烟雾形成中非常活泼。

(4) Reaction of Aromatic Hydrocarbons with HO· :

(4) 芳烃与 HO· 的反应：

(5) Aldehyde reactions.

(5) 醛反应。

1) With HO· :

1) 和 HO· :

$$R-\overset{O}{\overset{\|}{C}}-H + HO\cdot + O_2 \longrightarrow R-\overset{O}{\overset{\|}{C}}-OO\cdot + H_2O$$

$$\underset{H}{\overset{H}{>}}C=O + HO\cdot + \frac{3}{2}O_2 \longrightarrow CO_2 + HOO\cdot + H_2O$$

(2) Photochemical：

(2) 光化学：

$$R-\overset{O}{\overset{\|}{C}}-H + h\nu + 2O_2 \longrightarrow ROO\cdot + CO + HOO\cdot$$

$$\begin{matrix}H\\H\end{matrix}C=O + h\nu + 2O_2 \longrightarrow CO + 2HOO\cdot$$

(6) Sequence of reactions leading to photochemically active NO_2.

1) Key to smog-forming process:

$$RH + HO\cdot \longrightarrow R\cdot + H_2O$$
$$R\cdot + O_2 \longrightarrow ROO\cdot$$
$$ROO\cdot + NO \longrightarrow RO\cdot + NO_2$$

2) Chain reactions re-initiated by:

$$NO_2 + h\nu \longrightarrow NO\cdot + O$$
$$RH + O \longrightarrow R\cdot + HO\cdot$$

(7) Peroxyacyl nitrate formation:

$$R-\overset{O}{\underset{\|}{C}}-OO\cdot + NO_2 \longrightarrow R-\overset{O}{\underset{\|}{C}}-OO\cdot + NO_2$$

When R is CH_3, peroxyacetyl nitrate is the product.

(8) Peroxyacyl nitrates are significant air pollutants.

1) Characteristic of photochemical smog.

2) Eye irritants and mutagens.

3) Potent phytotoxins that adversely affect plants.

(9) Formation of alkyl nitrates and nitrites:

1) $RO\cdot + NO_2 \rightarrow RONO_2$.

2) $RO\cdot + NO \rightarrow RONO$.

(10) Nitrate Radical.

1) NO_3 is an important species in smog formation, especially at night.

2) Rapid photodissociation in daylight:

$$NO_3 + h\nu(\lambda < 700nm) \longrightarrow NO + O_2$$
$$NO_3 + h\nu(\lambda < 580nm) \longrightarrow NO_2 + O$$

3) NO_3 rapidly adds across double bonds in alkenes.

(6) 导致光化学活性强的 NO_2 产生的反应顺序。

1) 烟雾形成过程的关键：

2) 链反应由以下反应重新启动：

(7) 过氧酰基硝酸酯的形成：

当 R 为 CH_3 时，过氧乙酰硝酸酯为产物。

(8) 过氧酰基硝酸酯是重要的空气污染物。

1) 光化学烟雾的特征。

2) 眼睛刺激物和诱变剂。

3) 对植物有害的强力植物毒素。

(9) 烷基硝酸酯和亚硝酸异戊酯的形成：

1) $RO\cdot + NO_2 \rightarrow RONO_2$；

2) $RO\cdot + NO \rightarrow RONO$。

(10) 硝酸盐自由基。

1) NO_3 是烟雾形成中的重要物种，尤其是在夜间。

2) 日光下快速光解离：

3) NO_3 在烯烃中通过双键快速加成。

(11) Photolyzable compounds in the atmosphere.

1) Most important is NO_2:

$$NO_2 + h\nu\,(\lambda<394\,nm) \longrightarrow NO + O$$

2) Photodissociation of carbonyls, especially formaldehyde:

$$CH_2O + h\nu\,(\lambda<335\,nm) \longrightarrow H\cdot + H\dot{C}O$$

3) Photodissociation of hydrogen peroxide:

$$HOOH + h\nu\,(\lambda<350\,nm) \longrightarrow HO\cdot + HO\cdot$$

4) Photodissociation of organic peroxides:

$$H_3COOH + h\nu\,(\lambda<350\,nm) \longrightarrow H_3CO\cdot + HO\cdot$$

(11) 大气中的可光解化合物。

1) 最重要的是 NO_2：

2) 羰基，特别是甲醛的光解离：

3) 过氧化氢的光解离：

4) 有机过氧化物的光解离：

重要名词和术语解析

Gross photochemical oxidant: In air-pollution parlance, **gross photochemical oxidant** is a substance in the atmosphere capable of oxidizing iodide ion to elemental iodine.

总光化学氧化剂：从空气污染的角度来看，**总光化学氧化剂**是一种能够将碘离子氧化为元素碘的物质。

2.5.6　Reactivity of Hydrocarbons
　　　碳氢化合物的反应性

知识框架

Reactivity based on speed of reaction with hydroxyl radical

(1) CH_4 least reactive, but still important in smog formation because of abundance.

(2) Benzene, ethene, and n-hexane examples of intermediate reactivity.

(3) β-pinene from conifer trees about 9000 times as reactive as methane.

(4) d-limonene from orange rind about 19000 times as reactive as methane.

基于与羟基自由基反应速度的反应性

(1) CH_4 反应性最低，但由于丰度高仍在烟雾形成中很重要。

(2) 苯、乙烯和正己烷是中间反应性的实例。

(3) 针叶树中的 β-蒎烯的活性是甲烷的9000倍。

(4) 来自橙皮的 d-柠檬烯的活性是甲烷的约19000倍。

2.5.7 Importance of HO$_x$/VOC Ratios
HO$_x$/VOC 比的重要性

知识框架

(1) Whether ground level ozone in smog is decreased by decreasing VOC or reducing HO$_x$ depends upon their relative abundances.

(2) A single HO· radical can produce numerous ozone molecules, a process of propagation.

(3) Radical propagation is stopped by termination processes.

(4) These processes and the production of ground-level ozone depend upon HO$_x$/VOC ratios.

(1) 是否能通过降低 VOC 或减少 HO$_x$ 来降低烟雾中的地面臭氧含量,取决于它们的相对丰度。

(2) 一个 HO· 自由基可产生大量臭氧分子,这是一个增长过程。

(3) 自由基增长通过终止过程而停止。

(4) 这些过程和地面臭氧的产生取决于 HO$_x$/VOC 比。

2.5.8 Inorganic Products from Smog
烟雾的无机产物

知识框架

Two major classes are sulfates and nitrates

(1) These inorganics contribute to:
1) Acidic precipitation.
2) Corrosion.
3) Reduced visibility.
4) Adverse health effects.

(2) Atmospheric sulfur from SO_2 emissions.
1) SO_2 rapidly oxidized in photochemical smog.
2) By oxidant compounds:
①O_3.
②NO_3.
③N_2O_5.
3) By radicals (especially hydroxyl):
①HO·.
②HOO·.

两大类是硫酸盐和硝酸盐

(1) 这些无机物有助于:
1) 酸性降水;
2) 腐蚀;
3) 降低能见度;
4) 对健康产生不利影响。

(2) 来自 SO_2 排放的大气硫。
1) 在光化学烟雾中迅速被氧化。
2) 被氧化剂化合物:
①O_3;
②NO_3;
③N_2O_5。
3) 被自由基(尤其是羟基):
①HO·;
② HOO·;

③ RO·.
④ ROO·.

(3) Formation of inorganic nitrates and nitric acid:

1) HO· +NO_2→HNO_3.
2) $H_2O+N_2O_5$→$2HNO_3$.

(4) Nitrates and HNO_3 are very damaging smog products:

1) Corrosive.
2) Toxic to plants.

2.5.9 Effects of Smog
烟雾的影响

知识框架

(1) Human health and comfort.
Especially respiratory effects of ozone.

(2) Damage to materials (such as ozone attack on rubber).

(3) Effects on the atmosphere.
Especially reduction of visibility.

(4) Toxicity to plants:

1) From ozone.
2) From organic oxidants such as peroxyacetyl nitrate.
3) Damage largely from ozone and organic oxidants.

2.6 The Endangered Global Atmosphere
面临危险的全球大气

2.6.1 Saving the Atmosphere to Save Ourselves
拯救大气，拯救我们自己

知识框架

Atmosphere crucial to life on Earth

(1) Atmosphere endangered, especially

③ RO·；
④ ROO·。

(3) 形成无机硝酸盐和硝酸：

1) HO· +NO_2→HNO_3；
2) $H_2O+N_2O_5$→$2HNO_3$。

(4) 硝酸盐和HNO_3是非常有害的烟雾产物：

1) 腐蚀性；
2) 对植物有毒。

(1) 人类健康与舒适。
特别是臭氧对呼吸的影响。

(2) 损坏材料（如臭氧侵蚀橡胶）。

(3) 对大气的影响。
特别是能见度的降低。

(4) 对植物的毒性。

1) 来自臭氧；
2) 来自有机氧化剂，如过氧乙酰硝酸盐；
3) 很大程度上受到臭氧和有机氧化剂的损害。

大气对地球生命至关重要

(1) 大气受到威胁，尤其

from carbon dioxide emissions causing global climate change.

May be reaching a tipping point.

(2) Atmosphere's close relationship with other environmental spheres.

(3) Preservation of the atmosphere's natural capital：

1) O_2 for respiration.
2) CO_2 for photosynthesis.
3) N_2 fixed for living matter.
4) Temperature stabilization.
5) Radiation filter.
6) Waste assimilation.

是二氧化碳排放引起的全球气候变化。

可能达到临界点。

(2) 大气与其他环境圈层的紧密关系。

(3) 保护大气的自然资本：

1) 氧气用于呼吸；
2) 二氧化碳用于光合作用；
3) 固定的氮气用于生物；
4) 稳定温度；
5) 过滤辐射；
6) 吸收废物。

2.6.2 the Earth's Evolving Atmosphere and Climate Change 地球不断演变的大气与气候变化

知识框架

(1) Lovelock's Gaia Hypothesis：Organisms have established and maintained the atmospheric O_2/CO_2 balance that has maintained Earth's climate and moist conditions conducive to life.

Human activities must avoid upsetting this balance.

(2) Earth's present oxygen-rich atmosphere was established by photosynthetic bacteria：

1) $CO_2 + H_2O + h\nu \rightarrow \{CH_2O\} + O_2$.
2) Evidence of O_2 production from iron oxide deposits.
3) $4Fe^{2+} + O_2 + 4H_2O \rightarrow 2Fe_2O_3 + 8H^+$.

(1) 拉夫洛克的盖亚假说：生物体已经建立并维持了大气中的 O_2/CO_2 平衡，从而维持了地球的有利于生命的气候和潮湿条件。

人类活动必须避免破坏这种平衡。

(2) 地球目前的富氧大气是由光合作用细菌建立的：

1) $CO_2 + H_2O + h\nu \rightarrow \{CH_2O\} + O_2$；
2) 从氧化铁沉积物获得产生 O_2 的证据；
3) $4Fe^{2+} + O_2 + 4H_2O \rightarrow 2Fe_2O_3 + 8H^+$。

(3) Accumulation of O_2 in the atmosphere enabled development of oxic organisms.

(4) Establishment of stratospheric ozone layer that absorbs ultraviolet radiation enables terrestrial life to exist.

(5) Organisms regulate atmospheric CO_2 and maintain climate.

(3) 大气中氧气的积累促进了有氧生物的发展。

(4) 产生吸收紫外线辐射的平流层臭氧层,使地球生命存在。

(5) 生物调节大气中的 CO_2 并维持气候。

2.6.3 Effects of the Anthrosphere on Atmosphere and Climate
人类活动圈对大气和气候的影响

知识框架

Human activities strongly influence the atmosphere particularly through emissions that:

(1) Warm the atmosphere by absorbing outgoing infrared.

(2) Pollutant particles that scatter and reflect sunlight.

(3) Photochemically reactive species such as NO_2.

(4) Emissions that catalyze destruction of stratospheric ozone.

人类活动强烈影响大气,特别是通过排放:

(1) 通过吸收外来红外线来温暖大气;

(2) 散射和反射阳光的污染物颗粒;

(3) 光化学反应性物质,例如 NO_2;

(4) 催化平流层臭氧破坏的排放。

In 1957 Revelle and Suess called human effects on the atmosphere a "massive geophysical experiment".

1957 年,Revelle 和 Suess 将人类对大气的影响称为"大规模地球物理实验"。

2.6.4 The Greatest Threat to the Atmosphere and the Earth System: Global Warming
对大气和地球系统的最大威胁:全球变暖

知识框架

(1) Evidence of massive changes in climate in past.

(2) Currently in 10000-year interglacial period called Holocene.

(1) 过去气候发生重大变化的证据。

(2) 目前处于 10000 年的间冰期,称为全新世。

(3) Concern over positive feedback mechanisms that could result in drastic climate change.

1) Example1: Warming melts ice cover that leads to greater absorption of light and more warming.

2) Example2: Warming releases methane gas from geosphere and sediments resulting in more infrared absorption and accelerated warming.

(4) Incoming solar energy in the visible region in a wavelength range of 400~800 nanometers with maximum intensity at a wavelength of about 500nm.

(5) Re-absorption of outbound infrared, especially by CO_2, has a warming effect.

(6) Equivalent amount of energy leaves the Earth system, primarily as infrared radiation with maximum intensity at about $10\mu m$, mostly within the range between $2\mu m$ and $40\mu m$.

(7) Global warming is caused by: reabsorption of outgoing infrared by CO_2, H_2O vapor, CH_4, other species in the atmosphere.

(8) Global warming is generally a good thing, but bad if it happens in excess.

(9) Concern over global warming from record hot years during the last two decades.

(10) Doubling of atmospheric CO_2 levels from pre-industrial levels by 2100 is projected to raise Earth's means surface temperature 1.5~4.5℃, which could have some very bad effects on climate:

(3) 对可能导致剧烈气候变化的积极反馈机制表示担忧。

1) 例1：变暖融化冰盖，导致更多的光线吸收和更多的加热。

2) 例2：变暖使甲烷气体从地层和沉积物中释放，从而导致更多的红外吸收和加速变暖。

(4) 可见光区域内400~800nm波长的入射太阳能，在约500nm波长处达到最大强度。

(5) 重新吸收反射的红外光，尤其是被CO_2吸收，具有增暖作用。

(6) 相当量的能量主要以红外辐射离开地球系统，最大强度约为$10\mu m$，大部分在$2\mu m$与$40\mu m$之间。

(7) 全球变暖原因：大气中的CO_2、H_2O蒸气、CH_4和其他物质对吸收的红外光进行再吸收。

(8) 一般来说，全球变暖是一件好事，但如果发生过多，将是一件坏事。

(9) 对过去二十年来创纪录的高温年份的全球变暖的担忧。

(10) 预计到2100年，大气中的CO_2水平将比工业化之前的水平翻一番，这将使地球的平均表面温度升高1.5~4.5℃，这可能会对气候产生非常不利的影响。

1) Record warm year in 2016.

2) Melting glaciers (important water source).

3) Melting polar ice.

4) Rising sea levels.

5) Expansion of subtropical arid regions and drought.

6) Lower ocean pH from dissolved CO_2 harming marine life.

Particles and Global Warming

(1) Effects of particles are complicated making their modeling difficult.

(2) Both cooling and warming effects:

1) Incoming visible radiation scattered, cooling effect.

2) Incoming radiation absorbed, warming effect.

3) Absorption of outbound infrared has warming effect.

(3) Sulfuric acid and sulfates from oxidation of SO_2 produces droplets containing water which has a generally cooling effect.

Methane and Other Greenhouse Gases

(1) Greenhouse gases other than CO_2 contribute to global warming.

1) Halocarbons.

2) Hydrohalocarbons.

3) N_2O.

(2) Methane, CH_4, is especially important.

1) Now at 1.8×10^{-6}.

2) Increasing about 0.2×10^{-7} per year.

(3) Methane from several important sources:

1) Natural gas leakage.

2) Coal mine emissions.

1) 2016年为创纪录的温暖的一年。

2) 冰川融化（重要的水源）。

3) 极地冰融化。

4) 海平面上升。

5) 扩大亚热带干旱地区和干旱。

6) 溶解 CO_2 降低海洋pH值，损害海洋生物。

颗粒物与全球变暖

(1) 粒子的影响十分复杂，因此很难建模。

(2) 既有降温又有增温作用:

1) 将传入的可见辐射分散，冷却作用;

2) 吸收入射辐射，增暖作用;

3) 吸收反射红外线具有增暖作用;

(3) SO_2 氧化产生的硫酸和硫酸盐产生的水滴含有水，通常具有冷却作用。

甲烷和其他温室气体

(1) 除 CO_2 以外的温室气体也导致全球变暖。

1) 卤代烃。

2) 氢卤代烃。

3) N_2O。

(2) 甲烷 CH_4 尤其重要。

1) 现在为 1.8×10^{-6}。

2) 每年增加约 0.2×10^{-7}。

(3) 甲烷来自几个重要来源:

1) 燃气泄漏;

2) 煤矿排放;

3) Petroleum recovery.
4) Burning savannas, tropical forests.
5) From methane entrapped in thawing permafrost.

(4) Much methane from biogenic sources including:
1) Degradation of organic matter such as in landfills.
2) Anoxic biodegradation in rice paddies.
3) Anoxic bacteria in ruminant animals.

(5) Radiative forcing of methane:
1) Absorption of infrared per molecule.
2) About 25 times that of CO_2.

3）石油采收；
4）燃烧稀树草原、热带森林；
5）融化永久冻土释放甲烷。

(4) 来自生物源的大量甲烷，包括：
1) 如垃圾填埋场中的有机物的降解；
2) 稻田中的缺氧生物降解；
3) 反刍动物中的缺氧细菌。

(5) 甲烷的辐射能力：
1) 每个分子吸收红外线；
2) 大约是 CO_2 的 25 倍。

2.6.5 Green Science and Technology to Alleviate Global Warming 缓解全球变暖的绿色科学技术

知识框架

(1) Minimization of CO_2 emissions.
1) Energy from sources other than carbon-based fuels.
2) Electricity from renewable and nuclear sources.
3) Hydrogen fuel.
4) More efficient use of carbon-based fuels.
5) Carbon sequestration.
6) Carbon taxes and cap-and-trade system.

(2) Counteracting measures.
1) Light-reflecting aerosols into atmosphere.
2) Probably not very practical.

(3) Adaptation.
1) Will be necessary.
2) Drought will require more efficient water utilization.

(1) 减少 CO_2 排放。
1) 碳基燃料以外的能源。
2) 可再生和核能来源的电力。
3) 氢燃料。
4) 更有效使用碳基燃料。
5) 碳封存。
6) 碳税和总量管制与交易制度。

(2) 应对措施。
1) 反射光的气溶胶进入大气。
2) 可能不太实用。

(3) 适应。
1) 将是必要的。
2) 干旱将需要更有效地利用水。

3) Fresh water from wastewater and saline water.

4) Abundant, renewable, non-carbon energy is key.

2.6.6　Acid Rain
酸雨

知识框架

(1) Acids stronger than CO_2 in aqueous precipitation.

1) H_2SO_4 (most common).

2) H_2SO_3.

3) HNO_3.

4) HCl.

5) Rain.

6) Mist.

7) Sleet.

8) Fog (may be especially acidic).

9) Acid rime (frozen cloudwater on surfaces).

(2) Acid deposition is a more general term.

1) Aqueous acids.

2) Acid gases (SO_2).

3) Acidic salts, such as NH_4HSO_4.

(3) Most acids in acid precipitation are secondary pollutants from oxidation:

1) $SO_2 \rightarrow H_2SO_4$.

2) $NO_x \rightarrow HNO_3$.

(4) Acid precipitation is a regional air pollution problem often spreading over several hundred kilometers.

1) Photochemical smog is a local air pollution problem typically over several tens of km.

2) Stratospheric ozone depletion is a global

(1) 降水溶液中的酸强于CO_2。

1) H_2SO_4（最常见）。

2) H_2SO_3。

3) HNO_3。

4) HCl。

5) 雨。

6) 薄雾。

7) 雨夹雪。

8) 雾（可能酸性特别强）。

9) 酸雾（表面冻结的云水）。

(2) 酸沉降是一个更通用的术语。

1) 液态酸。

2) 酸性气体（SO_2）。

3) 酸性盐，例如NH_4HSO_4。

(3) 酸沉降中的大多数酸是氧化的二次污染物：

1) $SO_2 \rightarrow H_2SO_4$；

2) $NO_x \rightarrow HNO_3$。

(4) 酸沉降是一个区域性空气污染问题，通常会扩散数百千米。

1) 光化学烟雾是当地的空气污染问题，通常超过几十千米。

2) 平流层臭氧消耗是一

air pollution problem.

Harmful Effects of Acid Precipitation

(1) Direct phytotoxicity to plants from strong acids.

(2) Phytotoxicity from acid-forming gases (SO_2, NO_x).

(3) Acidification of water bodies.

(4) Acidification of soil.

(5) Indirect phytotoxicity, especially from Al^{3+} in soil.

(6) Destruction of sensitive forests.

(7) Respiratory effects on humans and other animals.

(8) Corrosion of metals, such as in electrical relays.

(9) Deterioration of stone:
$$2H^+ + CaCO_3 \text{(limestone)} \longrightarrow Ca^{2+} + CO_2(g) + H_2O$$

(10) Associated effects.

(11) Visibility reduction and increased haziness from sulfate aerosols.

(12) Physical and optical properties of clouds.

(13) Potential reduction in greenhouse warming from increased cloud cover.

酸沉降的有害影响

(1) 强酸对植物的直接植物毒性。

(2) 产酸气体（SO_2，NO_x）的植物毒性。

(3) 水体酸化。

(4) 土壤酸化。

(5) 间接植物毒性，尤其是来自土壤中的 Al^{3+}。

(6) 破坏敏感森林。

(7) 对人类和其他动物的呼吸作用。

(8) 金属腐蚀，例如继电器。

(9) 石材变质：

(10) 相关效果。

(11) 硫酸盐气溶胶降低能见度并增加雾度。

(12) 云的物理和光学特性。

(13) 云量增加可能减少温室效应。

重要名词和术语解析

Acid rain: Precipitation made acidic by the presence of acids stronger than CO_2 (aq) is commonly called acid rain.

Acid deposition: the term applies to all kinds of acidic aqueous precipitation, including fog, dew, snow, and sleet. In a more general sense, acid deposition refers to the deposition on the earth's surface of aqueous acids, acid gases (such as SO_2), and acidic salts (such as NH_4HSO_4).

<u>酸雨</u>：由于存在比 CO_2（液相）更强的酸而使降水酸化，通常称为酸雨。

<u>酸沉降</u>：该术语适用于各种酸性水溶液沉降，包括雾、露、雪和雨夹雪。从更广泛的意义上讲，酸沉降是指酸、酸性气体（如 SO_2）和酸性盐（如 NH_4HSO_4）在地球表面的沉降。

Acid rime: Another source of precipitation heavy in the ammonium, sulfate, and nitrate ions associated with atmospheric acid is acid rime.

酸雾：酸雾是与大气酸相关的大量铵盐、硫酸盐和硝酸根离子沉降的另一个来源。

2.6.7 Stratospheric Ozone Destruction
平流层臭氧破坏

知识框架

(1) Ozone is produced in the stratosphere.
1) $O_2 + h\nu \rightarrow O + O$ ($\lambda < 242.4nm$).
2) $O + O_2 + M \rightarrow O_3 + M$ (energy-absorbing N_2 or O_2).

(2) Stratospheric ozone is destroyed by absorption of UV:
$$O_3 + h\nu \longrightarrow O_2 + O \ (\lambda < 325nm)$$
and by a series of reactions the net result of which is:
$$O_3 + O \longrightarrow O_2 + O_2$$

(3) About 350000 metric tons of ozone are formed and destroyed in the stratosphere daily.

(1) 臭氧在平流层中产生。
1) $O_2 + h\nu \rightarrow O + O$ ($\lambda < 242.4nm$)。
2) $O + O_2 + M \rightarrow O_3 + M$ (吸收能量的 N_2 或 O_2)。

(2) 平流层臭氧被紫外线吸收破坏：

通过一系列反应，最终结果是：

(3) 每天在平流层中形成并销毁约 35×10^4 t 的臭氧。

The Stratospheric Ozone Layer

(1) From about 15~35km in altitude:
1) Below 15km energetic UV required to split O_2 has been filtered out.
2) Above 35km there is very little oxygen and most is atomic O.

(2) If all stratospheric ozone were in a pure layer at 25℃ and 1atm❶ pressure, the ozone layer would be only 3mm thick.

平流层臭氧层

(1) 海拔约 15~35km。
1) 在 15km 以下，分解 O_2 所需的高能紫外线已被滤除。
2) 在 35km 以上，氧气很少，大部分是原子 O。

(2) 如果所有平流层臭氧都在 25℃ 和 1atm❶ 的压力下处于单纯的一层中，则臭氧层的厚度将仅为 3mm。

---

❶ 1atm = 101.325kPa。

(3) The total amount of ozone in the atmosphere above a particular point is measured in Dobson units, DU.

1) 1 DU = 0.001 atm-cm.

2) Average ozone layer thickness is 300DU.

Shielding Effect of the Ozone Layer

(1) Ozone absorbs ultraviolet radiation very strongly in the range 220~330nm.

(2) Effectively absorbs very damaging UV-B radiation in the range 290~320nm.

Essential protection for life on Earth's surface.

(3) Partial destruction of the stratospheric ozone layer could have adverse effects.

1) Increased skin cancer in humans.

2) Increased incidence of cataracts.

3) Damage to food crop plants.

4) Lowered productivity of marine phytoplankton.

Ozone Layer Destruction

(1) 1974 discovery of chlorofluorocarbon destruction of O_3.

Nobel Prize to Molina, Rowland, and Crutzen

(2) Stratospheric reactions of chlorofluorocarbons:

$$CF_2Cl_2 + h\nu \longrightarrow Cl\cdot + CF_2Cl\cdot$$

(3) Followed by catalytic destruction of ozone by Cl atoms:

1) $Cl\cdot + O_3 \rightarrow ClO\cdot + O_2$.

2) $ClO\cdot + O \rightarrow Cl\cdot + O_2$ (Cl atom to destroy more ozone).

（3）高于特定点的大气中臭氧的总量以 Dobson 单位（DU）测量。

（1）1DU = 0.001 atm-cm。

（2）平均臭氧层厚度为 300DU。

臭氧层的保护作用

（1）臭氧在 220~330nm 吸收紫外线能力非常强。

（2）有效吸收 290~320nm 范围内非常有害的 UV-B 辐射。

保护地球表面的生命。

（3）平流层臭氧层的部分破坏可能产生不利影响。

1）人类皮肤癌增加。

2）白内障发病率增加。

3）对粮食作物的损害。

（4）降低海洋浮游植物的生产力。

臭氧层破坏

（1）1974 年发现氯氟烃破坏 O_3。

诺贝尔奖获得者莫利纳、罗兰和克鲁岑。

（2）含氯氟烃的平流层反应：

（3）随后 Cl 原子催化破坏臭氧：

1）$Cl\cdot + O_3 \rightarrow ClO\cdot + O_2$；

2）$ClO\cdot + O \rightarrow Cl\cdot + O_2$（Cl 原子可破坏更多的臭氧）；

3) $O_3 + O \rightarrow O_2 + O_2$ (Net reaction up to 10000 times per Cl·).

An important reaction sequence involves formation and subsequent destruction of the dimer from ClO· to give Cl· that catalyzes O_3 destruction.

Antarctic ozone hole discovered in 1985

(1) Occurs in southern hemisphere springtime (September/October) when sunlight reaches stratospheric clouds in Antarctica.

(2) Active forms of Cl are stored in (largely crystalline ice) stratospheric Antarctic clouds during winter.

1) $ClONO_2$.

2) Cl_2.

3) HOCl.

(3) Reaction of $ClONO_2$ with HCl produces photochemically reactive Cl_2.

(4) Starting in September ultraviolet solar radiation reaches these clouds producing active Cl· that catalyzes ozone destruction.

1) $Cl_2 + h\nu \rightarrow Cl· + Cl·$.

2) $HOCl + h\nu \rightarrow HO· + Cl·$.

Green Chemistry Solutions to Stratospheric Ozone Depletion

(1) Restrictions from 1986 Montreal Protocol on Substances that Deplete the Ozone Layer.

1) Ban on production of chlorofluorocarbons (CFCs).

2) Now having noticeable effects on levels of ozone-depleting substances.

3) $O_3 + O \rightarrow O_2 + O_2$（每个 Cl·净反应多达 10000 次）。

一个重要的反应顺序是从 ClO·形成二聚体并随后对其进行破坏，从而生成催化 O_3 破坏的 Cl·。

1985 年发现南极臭氧洞

（1）在南半球的春季（9 月/10 月）发生，当阳光到达南极的平流层云时。

（2）冬季，Cl 的活性形式存储在（主要是冰晶）平流层南极云中。

1) $ClONO_2$。

2) Cl_2。

3) HOCl。

（3）$ClONO_2$ 与 HCl 反应生成具有光化学反应活性的 Cl_2。

（4）产生活性 Cl·，催化破坏臭氧。

1) $Cl_2 + h\nu \rightarrow Cl· + Cl·$。

2) $HOCl + h\nu \rightarrow HO· + Cl·$。

平流层臭氧消耗的绿色化学解决方案

（1）1986 年《蒙特利尔议定书》对消耗臭氧层物质的限制。

1) 禁止生产氯氟烃（CFC）。

2) 现在对消耗臭氧层物质的含量产生明显影响。

(2) Replacement of CFCs.

1) H-containing compounds destroyed in the troposphere by HO· radical.

2) HCFCs such as HCFC-22, $CHClF_2$ (still contain Cl).

3) HFCs such as HFC-134a, CH_2FCF_3 (no Cl, but greenhouse gas).

4) As of 2015, R-410A, a mixture of difluoromethane (CH_2F_2, called R-32) and pentafluoroethane (CHF_2CF_3 called R-125) had become popular.

5) Volatile hydrocarbons (popular in Europe but not allowed in U.S. because of flammability).

6) Carbon dioxide substitute for CFCs to make plastic foams.
①Insulation.
②Food containers.
③Packing.

（2）氯氟烃的替代品。

1) 在对流层被HO·自由基破坏的含H化合物。

2) HCFC，例如HCFC-22、$CHClF_2$（仍然含有Cl）。

3) HFC，例如HFC-134a、CH_2FCF_3（不含Cl，但有温室气体）。

4) 自2015年起，R-410A，一种由二氟甲烷（CH_2F_2，称为R-32）和五氟乙烷（CHF_2CF_3，称为R-125）的混合物开始流行。

5) 挥发性碳氢化合物（在欧洲很受欢迎，但由于易燃性在美国不被允许）。

6) 二氧化碳替代氯氟烃制造塑料泡沫。
①保温；
②食品容器；
③包装。

重要名词和术语解析

Hydrohaloalkane: Currently, the substitutes for ozone-destroying chlorofluorocarbon compounds are **hydrohaloalkanes**, compounds that contain at least one H atom.

氢卤代烷：目前，破坏臭氧的氯氟烃化合物的替代品是**氢卤代烷**，即至少包含一个H原子的化合物。

2.6.8 Atmospheric Brown Clouds
大气棕云

知识框架

(1) A layer of brown-hued air about 3km thick extending from the Arabian Peninsula across China and the western Pacific.

(2) Caused by inefficient burning practices.

（1）从阿拉伯半岛延伸到中国和西太平洋的大约3km厚的棕褐色空气层。

（2）由效率低下的燃烧行为造成。

1) Coal-fired power plants.
2) Slash-and-burn agriculture.
3) Cooking fires with wood or dung fuel.
(3) Adverse effects.
1) Darkening megacities such as Beijing, New Dehli.
2) Melting glaciers.
3) Decreased agricultural productivity.
4) Adverse health effects.

Yellow Dust
(1) Huge masses of windblown dust and sand.
1) Begin with winds over Mongolia and China spreading eastward.
2) Mix with brown cloud.
3) Associated with desertification and deforestation aggravated by global warming.
(2) Adverse economic effects.
Bad for clean manufacturing processes.
(3) Adverse health effects.
1) Asthma.
2) Lung disease.
3) Perhaps immune disease.
(4) The "good yellow dust".
1) Dust from Sahara blown to Amazon.
2) Fertilization of ocean, especially with iron.
3) Increased phytoplankton growth (base of ocean food chain).

2.6.9 Atmospheric Damage by Photochemical Smog
光化学烟雾对大气的损害

知识框架

(1) Urban atmosphere acting as a chemical reactor for:

1）燃煤电厂；
2）刀耕火种的农业；
3）用木头或粪便做饭。
（3）不利影响。
1）特大城市变暗，例如北京、新德里；
2）冰川融化；
3）农业生产力下降；
4）对健康的不良影响。

黄尘
（1）大量的风沙尘土。
1）从蒙古和中国的风向东扩散开始；
2）与棕云混合；
3）与全球变暖加剧的荒漠化和森林砍伐有关。
（2）不利的经济影响。
不利于清洁制造工艺。
（3）不良的健康影响。
1）哮喘；
2）肺部疾病；
3）可能造成免疫疾病。
（4）"好的黄尘"。
1）来自撒哈拉沙漠的灰尘吹到亚马孙；
2）海洋施肥，尤其是铁；
3）浮游植物增长（海洋食物链的基础）。

（1）城市大气充当化学反应器：

1) Hydrocarbons.
2) NO_x.
3) O_2.
4) Sulfur compounds.
(2) Driven by sunlight to produce:
1) Ozone.
2) Organic oxidants.
3) Aldehydes.
4) Organic particles.
5) Nitrates.
6) Sulfates.
7) Other.
(3) Photochemical smog is a secondary air pollutant.
(4) Forms in an air parcel about 1000m thick.
(5) Produces urban aerosol of particles.
1) Urban brown cloud.
2) Condensation aerosol particles that are quite small.
3) Most likely to scatter light.
4) Most respirable.
(6) Potentially carcinogenic aromatic hydrocarbon particles such as benzo(a)pyrene in smog.

Types of Damage from Photochemical Smog
(1) Adverse health effects.
Largely from ozone and other oxidants.
(2) Damage to crops.
1) Reduction of yields.
2) Poorer quality.
3) In extreme cases, crop destruction.

1) 碳氢化合物；
2) NO_x；
3) O_2；
4) 硫化合物；
(2) 受阳光驱动产生：
1) 臭氧；
2) 有机氧化剂；
3) 醛；
4) 有机颗粒；
5) 硝酸盐；
6) 硫酸盐；
7) 其他。
(3) 光化学烟雾是二次大气污染物。
(4) 在约1000m厚的空气包裹中形成。
(5) 产生颗粒城市气溶胶。
1) 城市棕云；
2) 小的冷凝气溶胶颗粒；
3) 最有可能散射光；
4) 最可呼吸。
(6) 潜在的致癌芳香烃颗粒，如烟雾中的苯并(a)芘。

光化学烟雾造成的损害类型
(1) 不良健康影响。
主要来自臭氧和其他氧化剂。
(2) 对农作物的损害。
1) 降低产量；
2) 降低质量；
3) 在极端情况下，农作物遭到破坏。

(3) Atmospheric damage.

Largely reduced visibility from urban aerosol.

(4) Damage to materials.

1) Effects of ozone and other oxidants on materials, especially cracking and deterioration of rubber.

2) Corrosion, especially from H_2SO_4 and HNO_3 produced by oxidation of acid gases.

(5) Photochemical smog may be observed even in lightly populated rural areas.

Burning of savanna grasses in rural areas produces NO_x and reactive hydrocarbons that produce smog.

2.6.10 The Urban Aerosol 城市气溶胶

知识框架

(1) The urban aerosol is a brown-cloud-type of phenomenon afflicting urban areas prone to photochemical smog.

1) Largely very small particles formed from gases.

2) Scatter light.

3) Most respirable.

4) Carry toxicants.

(2) Contain organic particles emitted directly or from reactions of organic gases.

1) End-products of photochemical smog.

2) PAH compounds.

(3) Inorganic products.

1) H_2SO_4.

(3) 大气破坏。

由于城市气溶胶,大大降低了能见度。

(4) 材料损坏。

1) 臭氧和其他氧化剂对材料的影响,尤其是橡胶的开裂和变质;

2) 腐蚀,尤其是由酸性气体氧化产生的 H_2SO_4 和 HNO_3 腐蚀。

(5) 甚至在人口稠密的农村地区,也可能会观察到光化学烟雾。

在农村地区燃烧稀树草原草会产生氮氧化物和反应性碳氢化合物,会产生烟雾。

(1) 城市气溶胶是棕云类型的现象,困扰着容易发生光化学烟雾的城市地区。

1) 由气体形成的非常小的微粒;

2) 散射光;

3) 最可呼吸;

4) 携带有毒物。

(2) 包含直接排放的或来自有机气体反应的有机颗粒。

1) 光化学烟雾的最终产物;

2) PAH 化合物。

(3) 无机产物。

1) H_2SO_4;

2) Sulfate salts.
3) HNO_3.
4) Nitrate salts.
5) NH_4 salts.
6) Acid fog aerosol.

2) 硫酸盐；
3) HNO_3；
4) 硝酸盐；
5) 铵盐；
6) 酸雾气溶胶。

重要名词和术语解析

<u>Urban aerosol</u>: The most visible manifestation of smog is the **urban aerosol**, which greatly reduces visibility in smoggy urban atmospheres.

<u>Acid fog</u>: Another kind of urban aerosol particulate matter of considerable concern is **acid fog**, which may have pH values below 2 due to the presence of H_2SO_4 or HNO_3.

城市气溶胶：烟雾的最明显表现是**城市气溶胶**，大大降低了烟雾弥漫的城市大气的能见度。

酸雾：酸雾是另一种引起关注的城市气溶胶微粒物质，由于存在H_2SO_4或HNO_3，**酸雾**的pH值可能低于2。

2.6.11　What can be done?
　　　　　我们能做什么？

知识框架

(1) Major disruptions of the atmosphere and climate have the greatest potential for catastrophic and irreversible environmental destruction.

(2) Population pressures and desire for higher material living standards place heavy burdens on Earth's support systems.

(3) Damage to the climate from:
1) Use of low-quality, high-sulfur, fossil fuels.
2) Slash-and-burn agriculture.
3) Deforestation.
4) Destructive agricultural practices.
5) Desertification.

Three Approaches to the Global Warming Problem

(1) Minimization by reducing greenhouse gas emissions.

(1) 大气和气候的重大破坏对灾难性的、不可逆转的环境破坏具有最大潜力。

(2) 人口压力和对更高物质生活水平的渴望给地球的支撑系统带来沉重负担。

(3) 对气候的破坏：
1) 使用低质量、高硫的化石燃料；
2) 刀耕火种的农业；
3) 森林砍伐；
4) 破坏性的农业做法；
5) 荒漠化。

解决全球变暖问题的三种方法

(1) 通过减少温室气体排放使其最小化。

1) Alternate energy sources.
2) Energy conservation.
3) Carbon tax.
4) Reforestation.
5) Grassland restoration.
(2) Counteracting measures:
1) Light-reflecting particles to upper atmosphere.
2) Unlikely to be practical.
(3) Adaptation (will be required).
1) Especially increased efficiency and flexibility in the distribution and use of water.
2) Water desalination.
3) Complete restoration and purification of wastewater to drinking water standards.
4) Aquifer recharge.
5) Reforestation.
6) Restoration of desert lands.

"Tie-in strategy" with measures that are sensible even if global warming turns out to be less severe than is now expected.

1) 替代能源；
2) 节能；
3) 碳税；
4) 植树造林；
5) 草地恢复。
(2) 应对措施：
1) 高空大气的反光粒子；

2) 不太实际。
(3) 适应（必须）。
1) 尤其是提高水分配和使用的效率和灵活性；

2) 水淡化；
3) 完全按照饮用水标准恢复和净化废水；

4) 含水层补给；
5) 植树造林；
6) 恢复荒漠土地。

即使全球变暖的程度不如现在预期的那样，采取"关联战略"也是明智的。

重要名词和术语解析

Minimization: Minimization by reducing emissions of greenhouse gases, switching to alternate energy sources, increasing energy conservation, and reversing deforestation.

Counteracting measure: Counteracting measures, such as injecting light-reflecting particles into the upper atmosphere.

Adaptation: particularly through increased efficiency and flexibility of the distribution and use of water, which might be in very short supply in many parts of the world as a consequence of greenhouse warming.

最小化：通过减少温室气体排放、切换为替代能源、增加能源节约以及恢复被砍伐的森林来实现最小化。

应对措施：应对措施，如将反光颗粒注入高层大气。

适应：特别是通过提高水的分配和使用的效率和灵活性，由于温室变暖，在世界许多地方可能供不应求。

Exercises
习题

2-1 What phenomenon is responsible for the temperature maximum at the boundary of the stratosphere and the mesosphere?

2-2 What function does a third body serve in an atmospheric chemical reaction?

2-3 Why does the lower boundary of the ionosphere lift at night?

2-4 Considering the total number of electrons in NO_2, why might it be expected that the reaction of a free radical with NO_2 is a chain-terminating reaction?

2-5 It may be argued that wind energy, which is now used by growing numbers of large turbines to generate renewable electricity, is actually a form of solar energy. Explain on the basis of meteorological phenomena the rationale for this argument.

2-6 Suppose that 22.4 liters of air at STP is used to burn 1.50g of carbon to form CO_2, and that the gaseous product is adjusted to STP. What is the volume and the average molar mass of the resulting mixture?

2-7 If the pressure is 0.01atm❶ at an altitude of 38km and 0.001 at 57km, what is it at 19km (ignoring temperature variations)?

2-1 平流层和中间层边界的温度最大值是由什么现象造成的?

2-2 第三体在大气化学反应中起什么作用?

2-3 为什么电离层的下边界在夜间会抬升?

2-4 考虑到 NO_2 的电子总数,为什么会认为自由基与 NO_2 的反应是链终止反应?

2-5 有人认为,现在越来越多的大型涡轮机用来生产可再生电力的风能,实际上是太阳能的一种形式。请根据气象现象解释这种说法的根据。

2-6 假设用 22.4L 空气在 STP 下燃烧 1.50g 碳生成 CO_2,并将气态产物调节到 STP。所得混合物的体积和平均摩尔质量是多少?

2-7 如果在 38km 的高度上的压力是 0.01atm❶,在 57km 的高度上的压力是 0.001atm❶,那么在 19km 的高度上的压力是多少(忽略温度变化)?

---

❶ 1atm = 101.325kPa。

2-8 Measured in μm, what are the lower wavelength limits of solar radiation reaching the earth; the wavelength at which maximum solar radiation reaches the earth; and the wavelength at which maximum energy is radiated back into space?

2-9 Of the species O, HO*·, NO$_2^*$, H$_3$C·, and N$^+$, which could most readily revert to a nonreactive, "normal" species in total isolation?

2-10 Of the gases neon, sulfur dioxide, helium, oxygen, and nitrogen, which shows the most variation in its atmospheric concentration?

2-11 A 12.0-liter sample of air at 25℃ and 1.00atm pressure was collected and dried. After drying, the volume of the sample was exactly 11.50L. What was the percentage by mass of water in the original air sample?

2-12 The sunlight incident upon a 1 square meter area perpendicular to the line of transmission of the solar flux just above the Earth's atmosphere provides energy at a rate most closely equivalent to (   ).

A. that required to power a pocket calculator

B. that required to provide a moderate level of lighting for a 40-person capacity classroom illuminated with fluorescent lights

C. that required to propel a 2500 pound automobile at 55mph

2-8 以微米为单位，太阳辐射到达地球的波长下限是多少；太阳辐射最大限度到达地球的波长是多少；最大能量辐射回太空的波长是多少？

2-9 在O、HO*·、NO$_2^*$、H$_3$C·、N$^+$等物种中，哪种最容易在完全分离的情况下恢复为非反应性的"正常"物种？

2-10 在氖气、二氧化硫、氦气、氧气和氮气中，哪种气体的大气浓度变化最大？

2-11 收集并干燥25℃、1.00atm压力的12.0L空气样品。干燥后，样品的体积正好为11.50 L。原始空气样品中水的质量百分比是多少？

2-12 地球大气层正上方垂直于太阳通量传输线的1m^{2}区域内的太阳入射光，可提供的能量最接近于（   ）。

A. 为袖珍计算器供电

B. 具有荧光灯照明的可容纳40人的教室提供中等水平的照明

C. 以55mph❶的速度推动2500磅❷的汽车

---

❶ mph 是 mile per hour 的缩写，表示英里/小时，1mph＝1.609344km/h。

❷ 1pound（磅）＝0.454kg。

D. that required to power a 100-watt incandescent light bulb

E. that required to heat a 40-person classroom to 70°F when the outside temperature is −10°F

2-13 At an altitude of 50km, the average atmospheric temperature is essentially 0℃. What is the average number of air molecules per cubic centimeter of air at this altitude?

2-14 What is the distinction between chemiluminescence and luminescence caused when light is absorbed by a molecule or atom?

2-15 State two factors that make the stratosphere particularly important in terms of acting as a region where atmospheric trace contaminants are converted to other, chemically less reactive, forms.

2-16 What two chemical species are most generally responsible for the removal of hydroxyl radical from the unpolluted troposphere?

2-17 What is the distinction between the symbols "*" and "·" in discussing chemically active species in the atmosphere?

2-18 Of the following the true statement is (　　).

A. incoming solar energy is primarily in the form of infrared radiation

B. the very cold tropopause layer at the top of the troposphere is the major absorber of harmful ultraviolet radiation from the sun

D. 为100W的白炽灯泡供电

E. 当室外温度为−10°F❶时，将40人的教室加热到70°F

2-13 在50km的高度上，平均大气温度基本上为0℃。在此高度，每立方厘米空气平均有多少空气分子？

2-14 化学发光和光被分子或原子吸收后引起的发光有什么区别？

2-15 请说明使平流层成为大气中微量污染物向其他化学反应性较低的形式转化特别重要的区域的两个因素。

2-16 一般来说，哪两种化学物种可以清除未污染的对流层中的羟基自由基？

2-17 在讨论大气中的化学活性物种时，符号"*"和"·"的区别是什么？

2-18 下列说法中正确的是（　　）。

A. 入射的太阳光能主要以红外辐射的形式存在

B. 非常寒冷的对流层顶部是太阳有害紫外线辐射的主要吸收者

---

❶ 1°F = −17.22℃。

C. the stratosphere is defined as a region of the atmosphere in which temperature decreases with increasing altitude

D. a large fraction of solar energy is converted to latent heat by evaporation of water to produce water vapor in the atmosphere

E. temperature inversions are most useful because they cause air pollutants to disperse

2-19 Of the following the true statement is ( ).

A. chemiluminescence refers to a chemical reaction that results from a molecule having absorbed a photon of light

B. O^* denotes an excited oxygen atom

C. O_2^* denotes a free radical

D. HO· is an insignificant species in the atmosphere

E. the longer the wavelength of incoming solar radiation, the more likely it is to cause a photochemical reaction to occur.

2-20 Match the following pertaining to classes of atmospheric chemical species.

(A) NO_2 (B) H_2S

(C) NH_4HSO_4 (D) O_2^*

(1) Reductant

(2) Corrosive substance

(3) Photochemically active species

(4) Of the species shown, most likely to dissociate without additional outside input

2-21 Free radicals do not or are not ( ).

A. have unpaired electrons

B. normally highly reactive

C. last longer in the stratosphere than in the troposphere

D. take part in chain reactions

E. lose their energy spontaneously, reverting to a stable species by themselves.

2-22 Of the following the true statement is ( ).

A. the central feature of global weather is the redistribution of moisture from equatorial areas where it falls to polar areas where it freezes

B. cyclonic storms are caused by temperature inversions

C. temperature inversions limit the vertical circulation of air

D. albedo refers to the percentage of infrared radiation that is reabsorbed as energy is emitted from Earth

E. the troposphere has a homogeneous composition of all gases and vapors including water

2-23 Using numbers ranging from 1 to 4, put the following in order of their anticipated lifetime in the troposphere from the shortest-lived (1) to the longest-lived (4) and explain: CH_4, CCl_2F_2, NO_2^*, SO_2.

2-24 Earth's atmosphere is stratified into layers. Of the following, the true statement regarding this stratification, the characteristics of the layers, and the characteristics of species in the layers is ( ).

A. the stratosphere and troposphere have essentially the same composition

B. the upper boundary of the stratosphere

2-22 下列说法中正确的是（ ）。

A. 全球天气的核心特征是水分从赤道地区落到极地地区结冰重新分布

B. 气旋风暴是由逆温引起的

C. 逆温限制了空气的垂直循环

D. 反照率是指地球发出的红外辐射被重吸收为能量的百分比

E. 对流层中包括水在内的所有气体和蒸汽的成分都是均匀的

2-23 用1至4的数字，按下列物质在对流层中的预期寿命从最短的（1）到最长的（4）的顺序排列，并加以解释 CH_4、CCl_2F_2、NO_2^*、SO_2。

2-24 地球的大气层是分层的。下列关于这种分层、各层的特点以及各层中物种的特点，说法正确的是（ ）。

A. 平流层和对流层的成分基本相同

B. 平流层的上界比对流

is colder than the upper boundary of the troposphere because the former is higher

C. ozone is most desirable near Earth's surface in the troposphere

D. the composition of the troposphere is characterized by both its high and uniform content of water vapor

E. the boundary between the troposphere and the stratosphere serves as a barrier to the movement of one of the important constituents of tropospheric air

2-25 In 2006 the U.S. Environmental Protection Agency proposed lowering the allowable $PM_{2.5}$ level to $35\mu g/m^3$. How many particles would this be in a cubic meter of air assuming that all the particles were spheres of a diameter of $2.5\mu m$ and had a density of exactly $1g/cm^3$?

2-26 For small charged particles, those that are $0.1\mu m$ or less in size, an average charge of 4.77×10^{-10} esu is normally assumed for the whole particle. What is the surface charge in esu/cm^2 for a charged spherical particle with a radius of $0.1\mu m$?

2-27 What is the settling velocity of a particle having a Stokes' diameter of $10\mu m$ and a density of $1g/cm^3$ in air at 1.00atm pressure and 0℃ temperature? (The viscosity of air at 0℃ is 170.8cP. The density of air under these conditions is 1.29g/L)

层的上界冷，是因为平流层更高

C. 臭氧在地球表面附近的对流层中最为理想

D. 对流层的组成特点是既含有大量的水蒸气，又含有均匀的水蒸气

E. 对流层与平流层之间的边界是对流层空气中重要成分之一运动的障碍

2-25 2006年，美国环境保护局提议将$PM_{2.5}$的允许水平降低到$35\mu g/m^3$。假设所有的颗粒都是直径为$2.5\mu m$的球体，密度正好是$1g/cm^3$，那么$1m^3$的空气中会有多少颗粒？

2-26 对于小型带电粒子，即那些尺寸为$0.1\mu m$或更小的粒子，通常假设整个粒子的表面平均电荷为4.77×10^{-10} esu❶。半径为$0.1\mu m$的带电球体的表面电荷量是多少（esu/cm^2）？

2-27 斯托克斯直径为$10\mu m$、密度为$1g/cm^3$的颗粒在1.00atm❷和0℃温度的空气中的沉降速度是多少？（0℃时空气的黏度为170.8cP❸。在此条件下，空气的密度为1.29g/L）

---

❶ electrostatic units (esu)，静电单位制。

❷ 1atm=101.325kPa。

❸ $1cP = 10^{-3}Pa \cdot s$。

2-28 A freight train that included a tank car containing anhydrous NH_3 and one containing concentrated HCl was wrecked, causing both of the tank cars to leak. In the region between the cars a white aerosol formed. What was it, and how was it produced?

2-29 Examination of aerosol fume particles produced by a welding process showed that 2% of the particles were greater than $7\mu m$ in diameter and only 2% were less than $0.5\mu m$. What is the mass median diameter of the particles?

2-30 What two vapor forms of mercury might be found in the atmosphere?

2-31 Analysis of particulate matter collected in the atmosphere near a seashore shows considerably more Na than Cl on a molar basis. What does this indicate?

2-32 What type of process results in the formation of very small aerosol particles?

2-33 Which size range encompasses most of the particulate matter mass in the atmosphere?

2-34 Why are aerosols in the $0.1\sim1\mu m$ size range especially effective in scattering light?

2-35 Per unit mass, why are smaller particles relatively more effective catalysts for atmospheric chemical reactions?

2-36 In terms of origin, what are the three major categories of elements found in atmospheric particles?

2-37 What are the five major classes of

2-28 一列货运列车，包括一节装有无水 NH_3 的罐车和一节装有浓 HCl 的罐车发生事故，导致两节罐车泄漏。在车厢之间的区域形成了一种白色气溶胶。它是什么，是如何产生的？

2-29 对焊接过程中产生的气溶胶烟尘颗粒的检查表明，2%的颗粒直径大于$7\mu m$，只有2%的颗粒直径小于$0.5\mu m$。颗粒的质量中值直径是多少？

2-30 在大气中可能发现汞的哪两种蒸气形式？

2-31 对海岸附近大气中收集到的颗粒物的分析表明，按摩尔计算，Na 比 Cl 多得多。这说明了什么？

2-32 什么类型的过程会导致极小气溶胶颗粒的形成？

2-33 哪种尺寸范围包含了大气中大部分的颗粒物质量？

2-34 为什么$0.1\sim1\mu m$尺寸范围内的气溶胶对光的散射特别有效？

2-35 每单位质量，为什么较小的颗粒对大气化学反应的催化效果相对较好？

2-36 从来源来看，大气粒子中的可找到的元素主要有哪三类？

2-37 构成大气气溶胶颗粒的五

material making up the composition of atmospheric aerosol particles?

2-38 The size distribution of particles emitted from coal-fired power plants is bimodal. What are some of the properties of the smaller fraction in terms of potential environmental implications?

2-39 Of the following, the statement that is untrue regarding particles in the atmosphere is (   ). (explain)

A. dispersion aerosol particles formed by grinding up bulk matter are typically relatively large

B. very small particles tend to be acidic and often originate from gases

C. Al, Fe, Ca, and Si in particles often come from soil erosion

D. carcinogenic polycyclic aromatic hydrocarbons may be synthesized from saturated hydrocarbons under oxygen-deficient conditions

E. larger particles are more harmful because they contain more matter

2-40 Of the following, the species that is least likely to be a constituent of solid or liquid atmospheric particulate matter is (   ). (explain)

A. C
B. O_3
C. H_2SO_4
D. NaCl
E. benzo(a)pyrene

2-41 Of the following, the one that is not a characteristic of dispersion aerosols is

2-38 燃煤电厂排放的颗粒的大小分布是双峰型的。就潜在的环境影响而言，较小的部分有哪些特性？

2-39 下列各项中，关于大气中的粒子的说法不正确的是（   ）。（请解释）

A. 研磨大物质形成的分散气溶胶颗粒通常比较大

B. 极小的颗粒往往是酸性的，往往来源于气体

C. 颗粒中的 Al、Fe、Ca 和 Si 往往来源于土壤侵蚀

D. 致癌的多环芳烃可在缺氧条件下由饱和烃合成的

E. 较大的颗粒由于含有较多的物质，危害较大

2-40 下列物种中，最不可能成为固体或液体大气颗粒物成分的是（   ）。（请解释）

A. C
B. O_3
C. H_2SO_4
D. NaCl
E. 苯并(a)芘

2-41 在下列条件中，不是分散气溶胶的特征的是

( )（explain）

A. they are most readily carried into the alveoli of lungs

B. they are usually above 1μm in size

C. they are relatively easier to remove

D. they are generally less respirable

E. they are produced when bulk materials (larger particles) are ground up or sub-divided

2-42 Match the constituent of particulate matter from the left with its most likely source from the right, below.

(A) Si (B) PAH

(C) SO_3^{2-} (D) Pb

(1) Natural sources, soil erosion

(2) Incomplete combustion of hydrocarbons

(3) Element largely introduced by human activities

(4) Reaction of a gas in the atmosphere

2-43 Of the following, the most likely to be formed by pyrosynthesis is ( ). (explain)

A. Sulfate particles

B. Ammonium particles

C. Sulfuric acid mist

D. PAHs

E. Ozone in smog

2-44 Match each particle constituent below, left, with its likely source:

(A) Si

(B) V

(C) Benzo(a)pyrene

(D) Sulfuric acid droplets

( )。（请解释）

A. 最容易被带入肺泡

B. 通常尺寸超过 1μm

C. 相对较容易去除

D. 通常不易呼吸

E. 当将大物料（较大的颗粒）磨碎或细分时产生

2-42 将左边的颗粒物成分与右边的最可能的来源相匹配。

(A) Si (B) PAH

(C) SO_3^{2-} (D) Pb

(1) 自然源、土壤侵蚀

(2) 碳氢化合物的不完全燃烧

(3) 主要由人类活动引入的元素

(4) 气体在大气中的反应

2-43 下列各项中，最可能由热解作用形成的是（ ）。（请解释）

A. 硫酸盐颗粒

B. 铵颗粒

C. 硫酸雾

D. 多环芳烃

E. 烟雾中的臭氧

2-44 将下面每个粒子成分与其可能的来源相匹配。

(A) Si

(B) V

(C) 苯并(a)芘

(D) 硫酸液滴

(1) From gases in the surrounding atmosphere

(2) From natural sources

(3) Combustion of certain kinds of fuel oil

(4) From incomplete combustion

2-45 Why is it that "highest levels of carbon monoxide tend to occur in congested urban areas at times when the maximum number of people are exposed?"

2-46 Which unstable, reactive species is responsible for the removal of CO from the atmosphere?

2-47 (　　) of the following fluxes in the atmospheric sulfur cycle is smallest.

A. Sulfur species washed out in rainfall over land

B. Sulfates entering the atmosphere as "sea salt"

C. Sulfur species entering the atmosphere from volcanoes

D. Sulfur species entering the atmosphere from fossil fuels

E. hydrogen sulfide entering the atmosphere from biological processes in coastal areas and on land

2-48 Of the following agents, the one that would not favor conversion of sulfur dioxide to sulfate species in the atmosphere is (　　).

A. ammonia

B. water

C. contaminant reducing agents

D. ions of transition metals such as manganese

E. sunlight

（1）来自周围大气中的气体

（2）来自天然气

（3）某些种类的燃料石油的燃烧

（4）来自不完全燃烧

2-45 为什么说"在拥堵的城市地区，一氧化碳含量最高的时候往往是最多人接触的时候"？

2-46 哪种不稳定的反应性物种负责清除大气中的CO？

2-47 在大气硫循环中，（　　）通量最小。

A. 陆地上降雨冲刷出来的硫物种

B. 硫酸盐以"海盐"的形式进入大气

C. 从火山中进入大气的硫物种

D. 从化石燃料中进入大气的硫物种

E. 从沿海地区和陆地的生物过程中进入大气的硫化氢

2-48 下列介质中，不利于大气中二氧化硫转化为硫酸盐类的是（　　）。

A. 氨

B. 水

C. 污染物还原剂

D. 锰等过渡金属离子

E. 太阳光

2-49 The air inside a garage was found to contain $10^{-3}\%$ CO by volume at standard temperature and pressure (STP). What is the concentration of CO in mg/L and by mass?

2-50 How many metric tons of $w(S) = 5\%$ coal would be needed to yield the H_2SO_4 required to produce a 3.00cm rainfall of pH = 2.00 over a 100km² area?

2-51 In what major respect is NO_2 a more significant species than SO_2 in terms of participation in atmospheric chemical reactions?

2-52 Assume that an incorrectly adjusted lawn mower is operated in a garage such that the combustion reaction in the engine is:

$$C_8H_{18} + \frac{17}{2}O_2 \longrightarrow 8CO + 9H_2O$$

If the dimensions of the garage are 5m×3m×3m, how many grams of gasoline must be burned to raise the level of CO in the air to 0.1% by volume at STP?

2-53 A 12.0-L sample of waste air from a smelter process was collected at 25℃ and 1.00atm❶ pressure, and the sulfur dioxide was removed. After SO_2 removal, the volume of the air sample was 11.50L. What was the percentage by mass of SO_2 in the original sample?

2-54 What is the oxidant in the Claus

2-49 在标准温度和压力 (STP) 下,车库内的空气按体积分数计算含有 $10^{-3}\%$ 的 $CO_°$ CO 的浓度是多少 (mg/L),按质量计是多少 (质量分数)?

2-50 要在 100km² 的区域内产生 pH=2.00 的 3.00cm 降雨量,所需的 H_2SO_4 需要多少吨煤 $[w(S)=5\%]$?

2-51 就参与大气化学反应而言,NO_2 在哪些主要方面比 SO_2 更重要?

2-52 假设一台调整不当的割草机在车库中运行,发动机的燃烧反应是:

如果车库的尺寸是 5m×3m×3m,那么在 STP 下,要将空气中的 CO 含量提高到 0.1%(体积分数),必须燃烧多少克汽油?

2-53 在 25℃、1.00atm❶ 压力下,收集某冶炼厂生产过程中产生的 12.0L 废气样品,除去 SO_2。除去 SO_2 后,空气样品的体积为 11.50L,原样品中 SO_2 的含量(质量分数)是多少?

2-54 克劳斯反应中的氧化剂是

❶ 1atm=101.325kPa。

reaction? What is the commercial product of this reaction?

2-55 Carbon monoxide is present at a level of 10^{-5} by volume in an air sample taken at 15°C and 1.00atm❶ pressure. At what temperature (at 1.00atm pressure) would the sample also contain $10mg/m^3$ of CO?

2-56 Assume that the wet limestone process requires 1 metric ton of $CaCO_3$ to remove 90% of the sulfur from 4 metric tons of coal containing $w(S) = 2\%$. Assume that the sulfur product is $CaSO_4$. Calculate the percentage of the limestone converted to calcium sulfate.

2-57 If a power plant burning 10000 metric tons of coal per day with 10% excess air emits stack gas containing 0.01% by volume of NO, what is the daily output of NO? Assume the coal is pure carbon.

2-58 How many cubic kilometers of air at 25°C and 1atm pressure would be contaminated to a level of $5 \times 10^{-5}\%$ NO_x from the power plant discussed in the preceding question?

2-59 Match the following pertaining to gaseous inorganic air pollutants.
(A) CO　　(B) O_3
(C) SO_2　(D) NO
(1) Produced in internal combustion engines as a precursor to photochemical smog formation.

2-55 在15℃和1.00atm❶下采集的空气样品中，一氧化碳的体积含量为10^{-5}。在什么温度下（在1.00atm下），样品中也会含有$10mg/m^3$的CO？

2-56 假设湿法石灰石工艺需要1t $CaCO_3$从4t $w(S)=2\%$的煤中脱除90%的S，假设硫产品为$CaSO_4$。计算石灰石转化为硫酸钙的百分比。

2-57 如果一个发电厂每天燃烧10000t煤，并有10%的过量空气，排放的烟囱气体按体积计算含有0.01%的NO，那么每天输出的NO是多少？假设煤是纯碳。

2-58 在25℃和1atm❶下，上一问题中讨论的发电厂会将多少立方公里的空气污染到$5\times10^{-5}\%$的氮氧化物水平？

2-59 匹配下列与气态无机空气污染物有关的内容。
(A) CO　　(B) O_3
(C) SO_2　(D) NO
(1) 在内燃机中产生，作为光化学烟雾形成的前体。

---

❶ 1atm = 101.325kPa。

(2) Formed in connection with photochemical smog.

(3) Not associated particularly with smog or acid rain formation, but of concern because of its direct toxic effects.

(4) Does not cause smog to form, but is a precursor to acid rain.

2-60 Of the following, the one that is not an inorganic pollutant gas is ( ). (explain)

A. Benzo(a)pyrene

B. SO_2

C. NO

D. NO_2

E. H_2S

2-61 Of the following, the statement that is not true regarding carbon monoxide in the atmosphere is ( ). (explain)

A. it is produced in the stratosphere by a process that starts with H abstraction from CH_4 by HO·

B. it is removed from the atmosphere largely by reaction with hydroxyl radical

C. it is removed from the atmosphere in part by its being metabolized by soil microorganisms

D. it has some natural, as well as pollutant, sources

E. at its average concentration in the global atmosphere, it is probably a threat to human health

2-62 What important photochemical property

(2) 与光化学烟雾有关而形成的。

(3) 与烟雾或酸雨的形成没有特别的联系，但由于其直接的毒性影响而引起关注。

(4) 不会导致烟雾的形成，但却是酸雨的前兆。

2-60 下列各项中，不属于无机污染气体的是（ ）。（请解释）

A. 苯并(a)芘

B. SO_2

C. NO

D. NO_2

E. H_2S

2-61 下列各项中，关于大气中一氧化碳的说法不正确的是（ ）。（请解释）

A. 它在平流层中产生的过程是从 HO· 从 CH_4 中提取 H 开始的

B. 它主要是通过与羟自由基反应而从大气中清除的

C. 它从大气中清除的部分原因是被土壤微生物代谢掉了

D. 它有一定的自然来源，也有污染物来源

E. 按其在全球大气中的平均浓度，它可能对人类健康构成威胁

2-62 羰基化合物与 NO_2 有什么共同的重要光化学

do carbonyl compounds share with NO_2?

2-63 Of the following, the statement that is untrue regarding air pollutant hydrocarbons is (　　). (explain)

A. although methane, CH_4, is normally considered as coming from natural sources, and may be thought of as a non-pollutant, human activities have increased atmospheric methane levels, with the potential for doing harm

B. some organic species from trees can result in the formation of secondary pollutants in the atmosphere

C. alkenyl hydrocarbons containing the C=C group have a means of reacting with hydroxyl radical that is not available for alkanes

D. the reactivities of individual hydrocarbons as commonly measured for their potential to form smog, vary only about ±25%

E. most non-methane hydrocarbons in the atmosphere are of concern because of their ability to produce secondary pollutants

2-64 Of the following regarding organic air pollutants, the true statement is (　　). (explain)

A. carbonyl compounds (aldehydes and ketones) are usually the last organic species formed during the photochemical oxidation of hydrocarbons

B. carboxylic acids (containing the $—CO_2H$ group) are especially long-lived and persistent in the atmosphere

性质?

2-63 下列各项中，关于大气污染物碳氢化合物的说法不正确的是（　　）。（请解释）

A. 虽然甲烷 CH_4 通常被认为是来自自然界，可以认为是一种无污染物质，但人类活动使大气中甲烷含量增加，有可能造成危害

B. 树木中的一些有机物可导致大气中形成二次污染物

C. 含有 C=C 基团的烯烃有一种与羟基反应的方式，而烷烃则没有这种方法

D. 通常测量单个碳氢化合物形成烟雾的可能性，其反应性只变化了约 ±25%

E. 大气中的大多数非甲烷碳氢化合物因其产生二次污染物的能力而受到关注

2-64 下列关于有机大气污染物的说法中，正确的是（　　）。（请解释）

A. 羰基化合物（醛类和酮类）通常是碳氢化合物光化学氧化过程中最后形成的有机物

B. 羧酸（含 $—CO_2H$ 基团）在大气中的寿命特别长，持久性特别强

C. chlorofluorocarbons, such as CCl_2F_2, are secondary pollutants

D. peroxyacetyl nitrate, PAN, is a primary pollutant

E. HFCs pose a greater danger to the stratospheric ozone layer than do CFCs

2-65 Of the following species, the one which is the least likely product of the absorption of a photon of light by a molecule of NO_2 is (　　).

A. O

B. a free radical species

C. NO

D. NO_2^*

E. N atoms

2-66 (　　) of the following statements is true.

A. RO· reacts with NO to form alkyl nitrates

B. RO· is a free radical

C. RO· is not a very reactive species

D. RO· is readily formed by the action of stable hydrocarbons and ground state NO_2

E. RO· is not thought to be an intermediate in the smog-forming process

2-67 Of the following species, the one most likely to be found in reducing smog is: ozone, relatively high levels of atomic oxygen, SO_2, PAN, PBN.

2-68 Why are automotive exhaust pollutant hydrocarbons even more damaging to the environment than their quantities would indicate?

C. 氯氟烃，如 CCl_2F_2，是二次污染物

D. 过氧乙酰硝酸酯，PAN，是一次污染物

E. HFCs 对平流层臭氧层的危害比 CFCs 更大

2-65 下列物种中，NO_2 分子吸收光子的产物可能性最小的是（　　）。

A. O

B. 自由基物种

C. NO

D. NO_2^*

E. N 原子

2-66 以下说法正确的是（　　）。

A. RO· 与 NO 反应生成烷基硝酸盐

B. RO· 是一种自由基

C. RO· 不是一种反应性很强的物种

D. RO· 在稳定的碳氢化合物和基态 NO_2 的作用下容易形成

E. RO· 不被认为是烟雾形成过程中的中间体

2-67 下列物种中，最容易出现还原性烟雾的是：臭氧、含量相对较高的原子氧、SO_2、PAN、PBN。

2-68 为什么汽车尾气污染物碳氢化合物对环境的危害比其数量所显示的还要大？

2-69 At what point in the smog-producing chain reaction is PAN formed?

2-70 Which of the following species reaches its peak value last on a smog-forming day? (　　)

　　A. NO
　　B. oxidants
　　C. hydrocarbons
　　D. NO_2

2-71 What is the main species responsible for the oxidation of NO to NO_2 in a smoggy atmosphere?

2-72 Give reasons why a turbine engine should have lower hydrocarbon emissions than an internal combustion engine.

2-73 What pollution problem does a lean mixture aggravate when employed to control hydrocarbon emissions from an internal combustion engine?

2-74 Why is a modern automotive catalytic converter called a "three-way conversion catalyst"?

2-75 What is the distinction between reactivity and instability as applied to some of the chemically active species in a smog-forming atmosphere?

2-76 Why might carbon monoxide be chosen as a standard against which to compare automotive hydrocarbon emissions in atmospheres where smog is formed? What are some pitfalls created by this choice?

2-77 What is the purpose of alumina in an automotive exhaust catalyst? What kind of material actually catalyzes the destruction of pollutants in the catalyst?

2-69 在产生烟雾的链反应中，PAN 在什么时候形成？

2-70 在烟雾形成日，(　　) 物质最后达到峰值？

　　A. NO
　　B. 氧化剂
　　C. 碳氢化合物
　　D. NO_2

2-71 在充满烟雾的大气中，将 NO 氧化成 NO_2 的主要物种是什么？

2-72 说明为什么涡轮发动机的碳氢化合物排放量应低于内燃机。

2-73 采用稀混合气来控制内燃机碳氢化合物的排放，会加重什么污染问题？

2-74 为什么现代汽车催化器被称为"三向转化催化剂"？

2-75 反应性和不稳定性在应用于形成烟雾的大气中的一些化学活性物种时有何区别？

2-76 为什么会选择一氧化碳作为比较烟雾形成的大气中汽车碳氢化合物排放的标准，这种选择会带来哪些隐患？

2-77 汽车尾气催化剂中氧化铝的作用是什么，催化剂中到底是什么样的材料能催化污染物的破坏？

2-78 Some atmospheric chemical reactions are abstraction reactions and others are addition reactions. Which of these applies to the reaction of hydroxyl radical with propane? With propene (propylene)?

2-79 How might oxidants be detected in the atmosphere?

2-80 Each of the following occurs during smog formation. Place each in order from the one that occurs first (denoted 1) to the one that occurs last (denoted 5) and explain your choice. (   )

(1) an alkyl peroxyl radical, ROO·, is produced

(2) particles in the atmosphere obscure visibility

(3) NO and another product are produced from NO_2

(4) NO reacts to produce NO_2

(5) an alkyl radical, R·, is produced from a hydrocarbon

2-81 Why is ozone especially damaging to rubber?

2-82 Show how hydroxyl radical, HO·, might react differently with ethylene, $H_2C=CH_2$, and methane, CH_4.

2-83 Name the stable product that results from an initial addition reaction of hydroxyl radical, HO·, with benzene.

2-84 Of the following, the true statement is (   ). (explain)

A. NO_2 is not involved in the processes

2-78 大气中的化学反应有些是提取反应，有些是加成反应。其中哪些适用于羟基与丙烷的反应，羟基与丙烯反应呢？

2-79 如何检测大气中的氧化剂？

2-80 在烟雾形成的过程中，下列每一种情况都会发生。请将每一种发生的顺序从最先发生的（表示为1）到最后发生的（表示为5）解释一下。（   ）

(1) 产生烷基过氧自由基 ROO·

(2) 大气中的微粒遮挡了能见度

(3) NO 和另一种产物由 NO_2 产生

(4) NO 反应产生 NO_2

(5) 烃类产生烷基自由基 R·

2-81 为什么臭氧对橡胶的危害特别大？

2-82 解释羟基自由基 HO· 与乙烯 $H_2C=CH_2$ 和甲烷 CH_4 可能发生的不同反应。

2-83 请说出羟基自由基 HO· 与苯发生起始加成反应后的稳定产物。

2-84 下列说法中，正确的是（   ）。（请解释）

A. NO_2 不参与启动烟雾

that initiate smog formation, only in those that tend to stop it

B. NO undergoes photodissociation to start the process of smog-formation

C. NO_2 can react with free radical species to terminate chain reactions involved in smog formation

D. once NO_2 has undergone photodissociation, there is not a mechanism in a smog-forming atmosphere by which it can be regenerated

E. NO_2 is the most phytotoxic (toxic to plants) species present in a smoggy atmosphere

2-85 Of the following, the true statement pertaining to hydrocarbon reactivity in smog-formation is (　　). (explain)

A. alkanes are more reactive than alkenes (olefins)

B. reactivity is based on reaction with HO·

C. methane, CH_4, is the most reactive hydrocarbon

D. terpenes, such as d-limonene, are unreactive

E. hydroxyl radical is classified as an unreactive hydrocarbon

2-86 Smog aerosol droplets are composed of organic matter surrounding a small inorganic (mineral matter) core. Suggest what this shows regarding the process by which these aerosols are formed. Is the organic portion of the aerosol likely to be

形成的过程，只参与趋向于阻止烟雾形成的过程

B. NO 经过光解作用启动烟雾形成的过程

C. NO_2 能与自由基物种反应，终止参与烟雾形成的链式反应

D. NO_2 一旦发生光解作用，在形成烟雾的大气中就不存在可以再生的机制

E. NO_2 是烟雾大气中存在的植物毒性（对植物有毒）最大的物种

2-85 下列各项中，关于碳氢化合物在烟雾形成中的反应性的说法，正确的是（　　）。（请解释）

A. 烷烃比烯烃反应性强

B. 反应性是基于与 HO· 的反应

C. 甲烷 CH_4 是反应性最强的碳氢化合物

D. d-柠檬烯等萜类化合物不具有反应性

E. 羟基属于不具有反应性的碳氢化合物

2-86 烟雾气溶胶液滴是由有机物围绕着一个小的无机物（矿物质）核心组成的。请说明这些气溶胶的形成过程。气溶胶的有机部分是否可能是纯碳氢化合物

pure hydrocarbon (explain)?

2-87 How do modern transportation practices contribute to the kinds of atmospheric problems discussed in this chapter?

2-88 What is the rationale for classifying most acid rain as a secondary pollutant?

2-89 Distinguish among UV-A, UV-B, and UV-C radiation. Why does UV-B pose the greatest danger in the troposphere?

2-90 How does the extreme cold of stratospheric clouds in Antarctic regions contribute to the Antarctic ozone hole?

2-91 How does the oxidizing nature of ozone from smog contribute to the damage that it does to cell membranes?

2-92 What may be said about the time and place of the occurrence of maximum ozone levels from smog with respect to the origin of the primary pollutants that result in smog formation?

2-93 What is meant by a "tie-in strategy"?

2-94 Of the following, the statement that is untrue is ( ). (explain)

A. acid rain is denoted by any precipitation with a pH less than neutral (pH=7.00)

B. acid may be deposited as acidic salts and acid gases, in addition to aqueous acids

C. acid rain is a regional air pollution problem as distinguished from local or global problems

D. carbon dioxide makes rainfall slightly acidic

(请解释)?

2-87 现代运输实践如何导致本章讨论的各种大气问题?

2-88 将大多数酸雨归为二次污染物的理由是什么?

2-89 区分 UV-A、UV-B 和 UV-C 辐射,为什么 UV-B 在对流层中构成最大危险?

2-90 南极地区平流层云的极端寒冷如何对南极臭氧空洞造成影响?

2-91 烟雾中臭氧的氧化性质如何导致其对细胞膜的损害?

2-92 从导致烟雾形成的一次污染物的角度,从烟雾中出现最大臭氧水平的时间和地点可能怎么样?

2-93 "关联策略"是什么意思?

2-94 在以下陈述中,不正确的说法是( )。(请解释)

A. 酸雨表示为任何 pH 低于中性(pH=7.00)的降水

B. 酸可能会以酸性盐和酸性气体的形式沉降

C. 酸雨是区域性空气污染问题,有别于本地或全球性问题

D. 二氧化碳使降雨略有酸性

E. acid rain is often associated with elevated levels of sulfate ion, SO_4^{2-}

2-95 Of the following related to greenhouse gases and global warming, the true statement is (   ). (explain)

A. levels of greenhouse gas methane are increasing by about 1ppm❶ per year in the atmosphere

B. per molecule, methane has a greater effect on greenhouse warming than does carbon dioxide

C. radiative forcing of CO_2 is about 25 times that of CH_4

D. carbon dioxide is the only gas considered significant as a cause of greenhouse warming

E. although models predict greenhouse warming, there is no evidence in recent years that it may have in fact begun

2-96 Of the following, the true statement pertaining to the "Antarctic ozone hole" is (   ) (explain)

A. it reaches its peak during the Antarctic summer

B. it does not involve chlorine species

C. it involves only species that occur in the gas phase

D. it does not involve ClO radical

E. it is related to species that are frozen in stratospheric cloud particles at very low temperature

2-97 Of the following, the one that is not an

2-95 在以下与温室气体和全球变暖有关的论述中，正确的说法是（   ）。（请解释）

A. 大气中的温室气体甲烷水平每年以大约1ppm❶的速度增长

B. 甲烷分子对温室变暖的影响比二氧化碳更大

C. CO_2 的辐射力大约是 CH_4 的25倍

D. 二氧化碳是唯一被认为是导致温室效应的重要原因的气体

E. 尽管模型预测温室效应，近年来没有证据表明它实际上已经开始

2-96 在以下情况中，与"南极臭氧空洞"有关的正确说法是（   ）。（请解释）

A. 在南极夏季达到顶峰

B. 不涉及氯物种

C. 仅涉及在气相中存在的物种

D. 它不涉及 ClO 自由基

E. 它与在极低温度下冻结在平流层云颗粒中的物质有关

2-97 在下列情况中，不受酸雨

---

❶ $1ppm = 1 \times 10^{-6}$。

effect of acid rain is (　　). (explain)

A. direct phytotoxicity (plant toxicity) from H^+

B. phytotoxicity from acid-forming gases, such as SO_2

C. phytotoxicity from liberated Al^{3+}

D. toxicity to fish fingerlings from acid accumulated in lakes

E. all the above are effects

2-98 Of the following, the one that is not a measure for decreasing adverse effects on global climate is (　　). (explain)

A. minimization, such as reducing emissions of greenhouse gases

B. counteracting measures, such as injection of light-reflecting particles into the upper atmosphere

C. replacement of nuclear energy with fossil energy

D. adaptation, such as more efficient irrigation

E. "tie-in" strategy

2-99 Of the following, the true statement pertaining to "the Endangered Global Atmosphere" is (　　). (explain)

A. atmospheric SO_2 may indirectly help reduce the greenhouse effect

B. atmospheric carbon dioxide levels are projected to decrease after the year 2010

C. "nuclear winter" is of concern primarily for the greenhouse effect

D. the major effect of volcanic eruptions

影响的是(　　)。(请解释)

A. 来自 H^+ 的直接植物毒性

B. 来自形成酸的气体(如 SO_2)的植物毒性

C. 释放的 Al^{3+} 引起的植物毒性

D. 湖泊中积累的酸对鱼种的毒性

E. 以上都是影响

2-98 在以下措施中，(　　)不是减少全球气候不利影响的措施。(请解释)

A. 最小化，如减少温室气体的排放

B. 抵消措施，如注入反光颗粒进入高层大气

C. 用化石能替代核能

D. 适应，如更有效的灌溉

E. "关联"策略

2-99 在以下情况中，与"处境危险的全球大气"有关的正确说法是(　　)。(请解释)

A. 大气中的 SO_2 可能间接地有助于减少温室效应

B. 预计 2010 年以后大气中的二氧化碳水平会降低

C. "核冬天"主要是由于温室效应引起的

D. 火山喷发的主要影响

is greenhouse warming
E. photochemical smog is primarily a global problem, not regional or local

是温室变暖
E. 光化学烟雾主要是全球性问题，而不是区域性或局部性问题

扫描二维码查看
习题答案

3 Environmental Chemistry of the Hydrosphere
水圈环境化学

3.1 The Hydrosphere and Water Chemistry
水圈和水化学

3.1.1 Water: An Essential Part of Earth's Natural Capital
水：地球自然资本的重要组成部分

知识框架

Science of Water

(1) Hydrology: Study of water.

(2) Limnology: Science of fresh water.

(3) Oceanography: Science of oceans.

水科学

(1) 水文学：水研究。

(2) 湖泊学：淡水科学。

(3) 海洋学：海洋科学。

Environmental and resource problems with water

(1) Too little water: Drought, global warming.

(2) Too much water: Floods.

(3) Waterborne diseases.

(4) Contaminated water, water pollutants.

水的环境和资源问题

(1) 水太少：干旱，全球变暖。

(2) 水过多：洪水。

(3) 水传播疾病。

(4) 受污染的水，水污染物。

Water chemistry must consider many factors

(1) Where water is found.

(2) Interactions with geosphere (especially groundwater).

水化学必须考虑许多因素

(1) 发现水的地方。

(2) 与地圈（特别是地下水）的相互作用。

(3) Organisms (especially microorganisms) in water.

(4) Human influences (interaction with anthrosphere).

(5) Water very much involved with chemical fate and transport in the Earth System.

(3) 水中的生物（尤其是微生物）。

(4) 人类的影响（与人类活动圈的相互作用）。

(5) 水与地球系统中的化学品的归宿和迁移息息相关。

重要名词和术语解析

Hydrology: The study of water.

Limnology: Limnology is the branch of the science dealing with the characteristics of fresh water including biological properties, as well as chemical and physical properties.

Oceanography: Oceanography is the science of the ocean and its physical and chemical characteristics. The chemistry and biology of the Earth's vast oceans are unique because of the ocean's high salt content, great depth, and other factors.

水文学：对水的研究。

湖泊学：湖泊学是处理淡水特性（包括生物学特性以及化学和物理特性）的科学分支。

海洋学：海洋学是关于海洋的科学及其物理和化学特征。由于海洋的高盐含量、深度和其他因素，地球广阔的海洋具有独特的化学和生物学特性。

3.1.2　Sources and Uses of Water
　　　　水的来源和用途

知识框架

(1) Groundwater is a major source of municipal water.

Only source in some areas.

(2) Rapid depletion of groundwater in some areas.

1) Especially problematic in California's Central Valley—the vegetable garden of the U.S.

2) Severe in many areas of Africa and Asia.

(1) 地下水是市政水的主要来源。

在某些地区是唯一来源。

(2) 某些地区的地下水迅速枯竭。

1) 在加利福尼亚的中央谷地（美国的菜园）尤其成问题。

2) 在非洲和亚洲许多地区严重。

(3) Measures to alleviate groundwater shortages, especially deeper wells.

1) Lowered water table, no more water can be pumped.

2) Intrusions of saline water, especially in coastal areas.

3) Irreversible damage to aquifer compaction, cannot be recharged.

4) Surface subsidence, even structural damage.

5) Depletion of surface water in lakes and rivers.

6) Contributes to desertification—productive land becomes desert.

(3) 缓解地下水短缺，特别是更深水井的措施。

1) 地下水位降低，无法再抽水。

2) 盐水的入侵，特别是在沿海地区。

3) 对含水层压实造成不可逆转的破坏，无法补给。

4) 表面沉降，甚至结构损坏。

5) 湖泊和河流中的地表水枯竭。

6) 促进荒漠化——生产土地变成沙漠。

3.1.3　H_2O：Simple Formula, Remarkable Molecule
　　　　H_2O：简单的化学式，杰出的分子

知识框架

(1) Characteristics of H_2O molecule:

1) Electron pairs as far apart as possible→angular configuration.

2) Non-bonded pairs of electrons form hydrogen bonds.

(2) Polarity and H-bonding give H_2O special properties.

1) Good solvent for polar and ionic species.

2) Unique heat/temperature/density behavior.

1) High dielectric constant.

2) High surface tension.

3) High heat capacity.

4) High heat of fusion.

5) High heat of evaporation.

(1) H_2O 分子的特征：

1) 电子对距离尽可能远→角度配置。

2) 非键电子对形成氢键。

(2) 极性和氢键赋予 H_2O 特殊的性能。

1) 极性和离子物质的良好溶剂。

2) 独特的热/温度/密度行为。

1) 高介电常数。

2) 高表面张力。

3) 高热容。

4) 高熔化热。

5) 高蒸发热。

6) Stratification of bodies of water.

(3) Transparency allows photosynthesis by algae, photosynthetic bacteria, plants in water.

(4) Overturn in fall tends to stir up bottom sediment.

6）水体分层。

（3）透明性允许藻类、光合细菌、水中的植物进行光合作用。

（4）秋季的倾覆往往会搅动底部的沉积物。

重要名词和术语解析

Hydrogen bonds: Hydrogen bonds are a special type of bond that can form between the hydrogen in one water molecule and the oxygen in another water molecule.

Surface water: Surface water occurs primarily in streams, lakes, and reservoirs.

Wetland: Wetlands are flooded areas in which the water is shallow enough to enable growth of bottom-rooted plants.

Estuary: Estuaries are arms of the ocean into which streams flow. The mixing of fresh and salt water gives estuaries unique chemical and biological properties.

Thermal stratification: During the summer a surface layer (**epilimnion**) is heated by solar radiation and, because of its lower density, floats upon the bottom layer, or hypolimnion. This phenomenon is called **thermal stratification**.

Thermocline: The shear-plane, or layer between epilimnion and hypolimnion, is called the **the rmocline**.

Overturn: This disappearance of thermal stratification causes the entire body of water to behave as a hydrological unit, and the resultant mixing is known as **overturn**.

氢键：氢键是一种特殊的键，可以在一个水分子中的氢与另一个水分子中的氧之间形成。

地表水：地表水主要在溪流、湖泊和水库中。

湿地：湿地是被水淹没的地区，水较浅，足以使根深蒂固的植物生长。

河口：河口是河流流入海洋的手臂。淡水和盐水的混合赋予河口独特的化学和生物学特性。

热分层：夏季，表层（**表水层**）被太阳辐射加热，由于密度较低，漂浮在底层（均温层）之上。这种现象称为**热分层**。

温跃层：剪切平面或表水层和均温层之间的层称为**温跃层**。

倾覆：这种热分层现象的消失导致整个水体表现为一个水文单元，因此产生的混合称为**倾覆**。

3.1.4 Life in Water (Biota)
水中生物（生物群）

知识框架

(1) Autotrophic organisms.

1) Utilize solar or chemical energy.

2) Synthesize complex biochemicals from simple inorganic compounds.

3) Photosynthetic aquatic algae are producers that make biomass from CO_2 and other inorganic compounds.

(2) Heterotrophic organisms metabolize organic materials.

Decomposers (reducers) break down material of biological origin.

(3) Productivity is the ability of organisms in a body of water to produce biomass that is the basis of the food chain.

Eutrophication caused by excess productivity

(4) Decay of excess biomass.

Consumption of oxygen

Main physical factors affecting aquatic life

(1) Temperature.

(2) Transparency.

(3) Turbulence.

Oxygen in water

(1) Dissolved oxygen, DO.

(2) Biochemical oxygen demand, BOD, from degradable substance.

(1) 自养生物。

1) 利用太阳能或化学能。

2) 由简单的无机化合物合成复杂的生化产物。

3) 光合水生藻类是利用 CO_2 和其他无机化合物生产生物质的生产者。

(2) 异养生物代谢有机物质。

分解者（还原剂）分解生物来源的物质。

(3) 生产力是水体中生物体产生生物质的能力，而生物质是食物链的基础。

生产力过高造成的富营养化。

(4) 过量生物质的腐烂。

消耗氧。

影响水生生物的主要物理因素

(1) 温度。

(2) 透明度。

(3) 湍流。

水中的氧气

(1) 溶解氧，DO。

(2) 来自可降解物质的生化需氧量 BOD。

重要名词和术语解析

Biota: The living organisms (**biota**) in an aquatic ecosystem may be classified as either autotrophic or heterotrophic.

Autotrophic: **Autotrophic** organisms utilize solar or chemical energy to fix elements from simple, nonliving inorganic material into complex life molecules that compose living organisms.

Producer: Algae are the most important autotrophic aquatic organisms because they are **producers** that utilize solar energy to generate biomass from CO_2 and other simple inorganic species.

Heterotrophic: **Heterotrophic** organisms utilize the organic substances produced by autotrophic organisms as energy sources and as the raw materials for the synthesis of their own biomass.

Decomposer: **Decomposers** (or **reducers**) are a subclass of the heterotrophic organisms and consist of chiefly bacteria and fungi, which ultimately break down material of biological origin to the simple compounds originally fixed by the autotrophic organisms.

Productivity: The ability of a body of water to produce living material is known as its **productivity**.

Eutrophication: Excessive productivity results in decay of the biomass produced, consumption of dissolved oxygen, and odor production, a condition called **eutrophication**.

Dissolved oxygen: Dissolved oxygen (DO) frequently is the key substance in determining the extent and kinds of life in a body of water.

生物群：水生生态系统中的生物体（**生物群**）可分为自养型或异养型。

自养生物：**自养生物**利用太阳能或化学能将简单、无生命的无机材料中的元素固定到组成生命有机体的复杂生命分子中。

生产者：藻类是最重要的自养水生生物，因为它们是利用太阳能从 CO_2 和其他简单无机物种中产生生物质的**生产者**。

异养：**异养**生物利用自养生物产生的有机物质作为能源和合成自身生物质的原料。

分解者：**分解者**（或还原者）是异养生物的亚类，主要由细菌和真菌组成，它们最终将生物来源的物质分解为最初由自养生物固定的简单化合物。

生产力：水体生产生物材料的能力被称为**生产力**。

富营养化：生产力过高会导致产生的生物质腐烂，溶解氧的消耗和气味的产生，这种状况称为**富营养化**。

溶解氧：溶解氧（DO）经常是决定水体生命程度和种类的关键物质。

Biochemical oxygen demand (BOD): refers to the amount of oxygen utilized when the organic matter in a given volume of water is degraded biologically.

生化需氧量（BOD）：指当给定水中的有机物被生物降解时所利用的氧气量。

3.1.5 Introduction to Aquatic Chemistry
水化学简介

知识框架

(1) Common chemical phenomena occur in water.
　1) Acid-base.
　2) Solubility.
　3) Oxidation-reduction.
　4) Complexation.
　5) Biochemical (oxidation-reduction).
(2) Aquatic systems are complicated, open, and dynamic.
　1) Solid phases.
　2) Gas phases.
　3) Organisms.
(3) Simplified models based upon equilibrium conditions.
Rates of processes (kinetics) also important.

(1) 水中常见化学现象。
　1) 酸碱；
　2) 溶解；
　3) 氧化还原；
　4) 络合；
　5) 生化（氧化还原）。
(2) 水生系统复杂、开放且动态。
　1) 固相；
　2) 气相；
　3) 生物。
(3) 基于平衡条件的简化模型。
过程速率（动力学）也很重要。

3.1.6 Gases in Water
水中的气体

知识框架

(1) O_2 for fish.
(2) CO_2 to support algal growth CO_2.

(1) 给鱼提供 O_2。
(2) 支持藻类生长。

Henry's law: The solubility of a gas in a liquid is directly proportional to the partial

亨利定律：气体在液体中的溶解度和与液体接触的气体

pressure of the gas in contact with the liquid.

Oxygen in water

(1) From air (20.95% O_2 on basis of dry air).

(2) 8.32mg/L O_2 in water in equilibrium with air at 25℃.

(3) Decreases with increasing temperature.

Oxygen consumed by biodegradation of biomass, {CH_2O}

(1) {CH_2O} +O_2→CO_2+H_2O。

(2) 8.3mg O_2 consumed by only 7.8mg {CH_2O}.

的分压成正比。

水中的氧气

(1) 来自空气(O_2占干燥空气的20.95%)。

(2) 25℃时，水中的8.32mg/L O_2与空气保持平衡。

(3) 随着温度升高而降低。

生物质{CH_2O}的生物降解消耗氧气

(1) {CH_2O} + O_2 → CO_2 + H_2O。

(2) 仅7.8mg {CH_2O}消耗8.3mg O_2。

重要名词和术语解析

<u>Henry's Law</u>: The solubilities of gases in water are calculated with **Henry's Law**, which states that the solubility of a gas in a liquid is proportional to the partial pressure of that gas in contact with the liquid.

<u>亨利定律</u>：气体在水中的溶解度是根据亨利定律计算的，**亨利定律**指出气体在液体中的溶解度与该气体与液体接触的分压成正比。

3.1.7 Water Acidity and Carbon Dioxide in Water
水中的酸度和二氧化碳

知识框架

(1) Acidity: Capacity to neutralize OH^-.

(2) Alkalinity: Capacity to neutralize H^+.

(3) HCO_3^- as acid releases H^+:
$$HCO_3^- \longrightarrow H^+ + CO_3^{2-}$$

(4) HCO_3^- as base accepts H^+:
$$HCO_3^- + H^+ \longrightarrow H_2O + CO_2$$

(1) 酸度：中和OH^-的能力。

(2) 碱度：中和H^+的能力。

(3) 作为酸释放H^+：

(4) HCO_3^-作为碱接受H^+：

(5) Strong acids such as HCl contribute to free mineral acid, a pollutant.

Acid mine water from H_2SO_4

(6) Hydrated metal ions as acids:
$$Al(H_2O)_6^{3+} \longrightarrow Al(H_2O)_5(OH)^{2+} + H^+$$

Carbon Dioxide in Water

(1) From air and as product of organic matter decay.

1) In water mainly as $CO_2(aq)$.

2) Relatively little as H_2CO_3.

(2) Acid-base equilibria for carbon dioxide species:

$$CO_2(aq) + H_2O \longrightarrow H^+ + HCO_3^-, \quad K_{a_1} = \frac{[H^+][HCO_3^-]}{[CO_2]} = 4.45 \times 10^{-7}, \quad pK_{a_1} = 6.35$$

$$HCO_3^- \longrightarrow H^+ + CO_3^{2-}, \quad K_{a_2} = \frac{[H^+][CO_3^{2-}]}{[HCO_3^-]} = 4.69 \times 10^{-11}, \quad pK_{a_2} = 10.33$$

Slight acidity of unpolluted rainwater

Air is 406 parts per million CO_2 (in year 2017).

(1) Water in equilibrium with this air has $[CO_2(aq)] = 1.328 \times 10^{-5}$ mol/L.

(2) $CO_2(aq) + H_2O \rightarrow H^+ + HCO_3^-$.

(3) *$[H^+] = [HCO_3^-]$, $K_{a_1} = \frac{[H^+][HCO_3^-]}{[CO_2]} = 4.45 \times 10^{-7}$, $pK_{a_1} = 6.35$。

Solving the above gives:
(1) $[H^+] = 2.48 \times 10^{-6}$ mol/L.
(2) pH = 5.61.

解以上方程式得出:
(1) $[H^+] = 2.48 \times 10^{-6}$ mol/L;
(2) pH = 5.61。

重要名词和术语解析

Acid-base: **Acid-base** phenomena in water involve loss and acceptance of H^+ ion. Many species act as **acids** in water by releasing H^+ ion, others act as **bases** by accepting H^+, and the water molecule itself does both.

Acidity: **Acidity** as applied to natural water and wastewater is the capacity of the water to neutralize OH^-; it is analogous to **alkalinity**, the capacity to neutralize H^+.

Distribution of species diagram: with pH as a master variable.

酸碱: 水中的酸碱现象涉及 H^+ 离子的失和得。许多物种通过释放 H^+ 离子而在水中充当**酸**,其他物种则通过接受 H^+ 充当**碱**,而水分子本身同时扮演这两种角色。

酸度: 应用于天然水和废水的**酸度**是水中和 OH^- 的能力;它类似于**碱度**,即中和 H^+ 的能力。

物种分布图: 以 pH 值为主要变量。

3.1.8 Alkalinity
碱度

知识框架

(1) Alkalinity is the capacity to accept H^+ ion.
1) Most commonly: $HCO_3^- + H^+ \rightarrow CO_2(aq) + H_2O$.
2) Other common contributors: OH^-, CO_3^{2-}.

(2) Alkalinity:
1) Buffers water pH.
2) Carbon source for algal growth.
3) Reacts with some water treatment chemicals.

(3) Complete formula for alkalinity applicable over wide pH range:

(1) 碱度是接受 H^+ 离子的能力。
1) 最常见的是: $HCO_3^- + H^+ \rightarrow CO_2(aq) + H_2O$;
2) 其他常见贡献者: OH^-, CO_3^{2-}。

(2) 碱度:
1) 缓冲水的 pH 值;
2) 藻类生长的碳源;
3) 与某些水处理化学品发生反应。

(3) 适用于广泛 pH 值范围的碱度的完整公式为:

$$[alk] = [HCO_3^-] + 2[CO_3^{2-}] + [OH^-] - [H^+]$$

(4) A common value to assume for alkalinity is:

$$1.00 \to 10^{-3} \text{mol/L}$$

(5) At pH 7, essentially all alkalinity is from HCO_3^-.

(6) At pH 10 significant contributions from CO_3^{2-}.

About half the carbon at pH 10 as at pH = 7.

(4) 碱度的一个常用值是：

(5) 在 pH 值为 7 时，基本上所有碱度都来自 HCO_3^-。

(6) OH^- 在 pH 值为 10 时，CO_3^{2-}、OH^- 有显著贡献。

在 pH 值为 10 时，碳约为 pH = 7 时的一半。

3.1.9 Calcium and Other Metal Ions in Water 水中的钙和其他金属离子

知识框架

(1) Exist as hydrated metal ions such as $Al(H_2O)_6^{3+}$.

(2) The +3 hydrated metal ions lose H^+ so are acidic +3.

(1) 以水合金属离子的形式存在，例如 $Al(H_2O)_6^{3+}$。

(2) 水合金属离子会损失 H^+，因此呈酸性。

Dissolved Carbon Dioxide and Calcium Carbonate Minerals

For water in equilibrium with air ($4 \times 10^{-4}\%$ CO_2) and $CaCO_3(s)$:

溶解的二氧化碳和碳酸钙矿物质

对于与空气（$4 \times 10^{-4}\%$ CO_2）和 $CaCO_3(s)$ 保持平衡的水：

$$[CO_2] = 1.309 \times 10^{-5} \text{mol/L}, \quad [Ca^{2+}] = 5.18 \times 10^{-4} \text{mol/L},$$
$$[HCO_3^-] = 1.04 \times 10^{-3} \text{mol/L}, \quad [H^+] = 5.63 \times 10^{-9} \text{mol/L},$$
$$[CO_3^{2-}] = 8.63 \times 10^{-6} \text{mol/L}, \quad pH = 8.25$$

重要名词和术语解析

Hydrated: Metal ions in water solution are present in forms such as the **hydrated** metal cation $M(H_2O)_x$.

Water hardness: Calcium ion, along with magnesium and sometimes iron (Ⅱ) ion, accounts for **water hardness**.

水合：水溶液中的金属离子以水合金属阳离子 $M(H_2O)_x$ 等形式存在。

水的硬度：钙离子，和镁以及二价铁离子构成水的硬度。

3.1.10 Complexation and Chelation
络合与螯合

知识框架

$$Fe(H_2O)_6^{2+} + CN^- \rightleftharpoons Fe(H_2O)_5CN^+ + H_2O$$

$$Fe(H_2O)_5CN^- + CN^- \rightleftharpoons Fe(H_2O)_4(CN)_2 + H_2O$$

The reactions above show:

(1) Complexation.

(2) A ligand (CN^-) binding to a metal ion, a reversible process.

(3) Formation of a complex (ion), or coordination compound.

(4) CN^- ion in the above example is a unidentate ligand.

Chelation occurs with chelating agents that can bind in more than one place.

Organometallic compounds are formed by irreversible binding of organic groups to metal atoms.

以上反应表明:

(1) 络合;

(2) 一个金属离子结合一个配体（CN^-），可逆过程；

(3) 形成络合物（离子）或配位化合物；

(4) 上例中的 CN^- 离子是单齿配体。

螯合剂可以在多个位置结合，发生螯合作用。

有机金属化合物是通过有机基团与金属原子的不可逆结合而形成的。

Complexation and chelation affect ligands and metals

(1) Oxidation-reduction, decarboxylation, hydrolysis, and biodegradation of ligands.

(2) Solubilization, precipitation, adsorption, distribution, transport, and effects of metal ions.

(3) Biochemical effects of metals including bioavailability, toxicity, uptake by organisms

络合和螯合影响配体和金属

(1) 配体的氧化还原、脱羧、水解和生物降解。

(2) 金属离子的溶解、沉淀、吸附、分布、迁移及其影响。

(3) 金属的生化作用，包括生物可利用性、毒性、有机物的吸收。

Occurrence of chelating agents

(1) Biocompounds such as hemoglobin.

(2) Metabolic products of organisms.

螯合剂的出现

(1) 生物化合物，如血红蛋白。

(2) 生物的代谢产物。

(3) Pollutants such as from metal plating wastes.

Chelating agents in water may facilitate algal growth by making soluble nutrient iron available.

(3) 金属电镀废料等污染物。

水中的螯合剂可通过使营养铁可溶来促进藻类生长。

重要名词和术语解析

Speciation: The properties of metals dissolved in water depend largely upon the nature of metal species dissolved in the water. Therefore, **speciation** of metals plays a crucial role in their environmental chemistry in natural waters and wastewaters.

Metal complexes: metals may exist in water reversibly bound to inorganic anions or to organic compounds as **metal complexes**.

Complexation: the species that binds with the metal ion is called a **ligand**, and the product in which the ligand is bound with the metal ion is a **complex**, **complex ion**, or **coordination compound**.

Chelation: A special case of complexation in which a ligand bonds in two or more places to a metal ion is called **chelation**.

Organometallic: In addition to being present as metal complexes, metals may occur in water as **organometallic** compounds containing carbon-to-metal bonds.

Unidentate ligand: it possesses only one site that bonds to a metal ion.

Chelating agent: A chelating agent has more than one atom that may be bonded to a central metal ion at one time to form a ring structure.

形态：溶于水中的金属的特性在很大程度上取决于溶于水中的金属种类的性质。因此，金属的**形态**在天然水和废水的环境化学中起着至关重要的作用。

金属络合物：金属可能以**金属络合物**的形式与水中的无机阴离子或有机化合物可逆结合。

络合：与金属离子结合的物质称为**配体**，配体与金属离子结合的产物为**配合物**、**配位离子**或配位化合物。

螯合：络合的一种特殊情况，其中配体在两个或多个位置与金属离子键合，称为**螯合**。

有机金属：除以金属络合物形式存在外，金属还可能以含有碳金属键的**有机金属化合物**形式存在于水中。

单齿配体：仅具有一个与金属离子键合的位点。

螯合剂：一种螯合剂具有一个以上的原子，可以一次键合到中心金属离子上以形成环结构。

3.1.11 Bonding and Structure of Metal Complexes
金属配合物的键合与结构

重要名词和术语解析

Coordination sphere: The ligands are said to be contained within the **coordination sphere** of the central metal atom.

配位球：配体包含在中心金属原子的配位球内。

Coordination number: The **coordination number** of a metal atom, or ion, is the number of ligand electron-donor groups that are bonded to it.

配位数：金属原子或离子的**配位数**是与其键合的配体电子给体基团的数量。

3.1.12 Calculations of Species Concentrations
物种浓度的计算

知识框架

(1) Stepwise formation constants when multiple ligands may bond to metal ions.

(2) Overall formation constants for two or more ligands bonding.

(1) 逐级稳定常数：多个配体键合到金属离子。

(2) 累积稳定常数：两个或多个配体键合。

重要名词和术语解析

Formation constant: The stability of complex ions in solution is expressed in terms of **formation constants**.

Stepwise formation constant: Stepwise formation constants (K expressions) representing the bonding of individual ligands to a metal ion.

Overall formation constant: overall formation constants (β expressions) representing the binding of two or more ligands to a metal ion.

稳定常数：溶液中配合物离子的稳定性用**稳定常数**表示。

逐步稳定常数：逐步稳定常数（K 表达式）代表各个配体与金属离子的键合。

累积稳定常数：累积稳定常数（β 表达式）表示两个或多个配体与金属离子结合。

3.1.13 Complexation by Deprotonated Ligands
去质子配体的络合

知识框架

(1) Textbook example of copper chelation by completely ionized EDTA.

(2) At pH=11 at excess EDTA at 200mg/L the concentration of unchelated EDTA is only:

$$[Cu^{2+}] = 2.3 \times 10^{-20} \text{mol/L}$$
$$[CuY^{2-}] = 7.9 \times 10^{-5} \text{mol/L}$$
$$\frac{[Cu^{2+}]}{[CuY^{2-}]} = 3.0 \times 10^{-14}$$

1) Vastly reduced free copper ion concentration in excess EDTA.
2) Influence chemical effects of Cu^{2+}.
3) Influence biochemical effects of Cu^{2+}.

(1) 完全电离的 EDTA 螯合铜的教科书示例。

(2) 在 pH=11 且过量 EDTA 为 200mg/L 时，未螯合 EDTA 的浓度仅为：

1) 过量 EDTA 中的游离铜离子浓度大；
2) 影响 Cu^{2+} 的化学作用；
3) 影响 Cu^{2+} 的生化作用。

3.1.14 Complexation by Protonated Ligands
质子化配体的络合

知识框架

Examples shown with NTA, a triprotic acid chelating agent (structural formula of anion of NTA):

NTA（一种三酸螯合剂）的示例（NTA 阴离子的结构式）：

$$H_3T \longrightarrow H^+ + H_2T^-, \quad K_{a_1} = \frac{[H^+][H_2T^-]}{[H_3T]} = 2.18 \times 10^{-2}, \quad pK_{a_1} = 1.66$$

$$H_2T^- \longrightarrow H^+ + HT^{2-}, \quad K_{a_2} = \frac{[H^+][HT^{2-}]}{[H_2T^-]} = 1.12 \times 10^{-3}, \quad pK_{a_2} = 2.95$$

$$HT^{2-} \longrightarrow H^+ + T^{3-}, \quad K_{a_3} = \frac{[H^+][T^{3-}]}{[HT^{2-}]} = 5.25 \times 10^{-11}, \quad pK_{a_3} = 10.28$$

Throughout the normal range of pH in natural waters the predominant unchelated NTA species is HT^-.

在天然水的正常 pH 范围内，未螯合的 NTA 物种主要为 HT^-。

3.1.15 Solubilization of Lead Ion from Solids by NTA
NTA 从固体中溶解铅离子

知识框架

(1) A major concern with the introduction of pollutant chelating agents into water is their ability to dissolve and transport heavy metals.

(2) Several examples are presented for the chelating agent NTA reacting with lead-containing solids under various conditions.

(3) An approach is shown in which the overall reaction is broken down into its constituent reactions. These are then added together to get the overall reaction, then their equilibrium constant expressions are multiplied to give the expression for the overall reaction.

(4) Assume a water sample contains 25mg/L of the trisodium NTA salt, $N(CH_2CO_2Na)_3$ formula mass 257 in equilibrium with solid $Pb(OH)_2$. Under these conditions:

1) Unchelated NTA is present as HT^{2-}.

2) The other possible soluble form of NTA is the lead chelate, PbT^-.

3) At pH=8.00, $[OH^-]=1.00×10^{-6}$ mol/L.

4) The solubilization reaction is:
$$Pb(OH)_2(s) + HT^{2-} \longrightarrow PbT^- + OH^- + H_2O$$

5) The degree of solubilization of lead is given by the ratio $[PbT^-]/[HT^{2-}]$.

(1) 将污染物螯合剂引入水中的主要问题是其溶解和迁移重金属的能力。

(2) 这里展示了螯合剂 NTA 在各种条件下与含铅固体反应的几个例子。

(3) 一种方法是将整个反应分解成其组成反应。然后将它们加在一起以获得总反应，然后将其平衡常数表达式相乘以给出总反应的表达式。

(4) 假设水样中含有 25mg/L 的 NTA 三钠盐，$N(CH_2CO_2Na)_3$ 分子质量 257，与固体 $Pb(OH)_2$ 处于平衡状态。在这些条件下：

1) 未螯合的 NTA 以 HT^{2-} 的形式存在；

2) NTA 的另一种可能的可溶形式是铅螯合物 PbT^-；

3) pH=为 8.00 时，$[OH^-]=1.00×10^{-6}$ mol/L。

4) 溶解反应是：

5) 铅的溶解度由 $[PbT^-]/[HT^{2-}]$ 之比给出。

6) $[PbT^-]/[HT^{2-}] = 20.7$ under the conditions given showing that essentially all the NTA is bound to lead as soluble PbT^-:

$$Pb(OH)_2(s) \rightleftharpoons Pb^{2+} + 2OH^-, \quad K_{sp} = [Pb^{2+}][OH^-]^2 = 1.61 \times 10^{-20}$$

$$HT^{2-} \rightleftharpoons H^+ + T^{3-}, \quad K_{a_3} = \frac{[H^+][T^{3-}]}{[HT^{2-}]} = 5.25 \times 10^{-11}$$

$$Pb^{2+} + T^{3-} \rightleftharpoons PbT^-, \quad K_f = \frac{[PbT^-]}{[Pb^{2+}][T^{3-}]} = 2.45 \times 10^{11}$$

$$H^+ + OH^- \rightleftharpoons H_2O, \quad \frac{1}{K_w} = \frac{1}{[H^+][OH^-]} = \frac{1}{1.00 \times 10^{-14}}$$

$$Pb(OH)_2(s) + HT^{2-} \rightleftharpoons PbT^- + OH^- + H_2O, \quad K = \frac{[PbT^-][OH^-]}{[HT^{2-}]} = \frac{K_{sp}K_{a_3}K_f}{K_w} = 2.07 \times 10^{-5}$$

$$\frac{[PbT^-]}{[HT^{2-}]} = \frac{K}{[OH^-]} = \frac{2.07 \times 10^{-5}}{1.00 \times 10^{-6}} = 20.7$$

Reaction of NTA with Lead Carbonate NTA 与碳酸铅的反应

(1) Lead carbonate, $PbCO_3$, is the predominant lead species in water under many conditions.

(1) 在许多情况下,碳酸铅 $PbCO_3$ 是水中的主要铅物种。

(2) The water containing 25mg Na_3T in equilibrium with $PbCO_3$ at pH = 7.00. The following apply:

(2) 在 pH = 7.00 的水中含有 25mg Na_3T 与 $PbCO_3$ 平衡。有以下反应:

1) Reaction:

1) 反应:

$$PbCO_3(s) + HT^{2-} \rightleftharpoons PbT^- + HCO_3^-, \quad K = \frac{[PbT^-][HCO_3^-]}{[HT^{2-}]} = \frac{K_{sp}K_{a_3}K_f}{K_{a_2}} = 4.06 \times 10^{-2}$$

2) Assume a reasonable value of $[HCO_3^-] = 1.00 \times 10^{-3}$ mol/L.

2) 假设 $[HCO_3^-] = 1.00 \times 10^{-3}$ mol/L;

3) Ratio of chelated to unchelated NTA:

3) 螯合的与未螯合的 NTA 的比例:

$$\frac{[PbT^-]}{[HT^{2-}]} = \frac{K}{[HCO_3^-]} = \frac{4.06 \times 10^{-2}}{1.00 \times 10^{-3}} = 40.6$$

This ratio shows that under typical conditions NTA would dissolve lead from solid $PbCO_3$.

(3) The reaction and its equilibrium constant of NTA with $PbCO_3(s)$ in the presence of 1.00×10^{-3} mol/L Ca^{2+} at pH = 7.00 and $[HCO_3^-] = 1.00 \times 10^{-3}$ mol/L, are

$$PbCO_3(s) + CaT^- + H^+ \rightleftharpoons Ca^{2+} + HCO_3^- + PbT^-, \quad K'' = \frac{[Ca^{2+}][HCO_3^-][PbT^-]}{[CaT^-][H^+]} = 5.24$$

Putting the known concentrations into this expression and solving:

1) $\dfrac{[PbT^-]}{[HCO_3^-]} = 0.524$.

2) About 1/3 of the NTA is bound with Pb^{2+} and 2/3 with Ca^{2+}.

3) Therefore, the presence of excess Ca^{2+} significantly inhibits NTA from dissolving lead from lead carbonate.

该比例表明,在典型条件下,NTA 会溶解固体 $PbCO_3$ 中的铅。

(3) 在 pH = 7.00 和 $[HCO_3^-] = 1.00 \times 10^{-3}$ mol/L 的条件下,在 1.00×10^{-3} mol/L Ca^{2+} 存在下,NTA 与 $PbCO_3$ 的反应是和其平衡常数为:

将已知浓度放入该表达式并求解:

1) $\dfrac{[PbT^-]}{[HCO_3^-]} = 0.524$;

2) 约 NTA 的 1/3 与 Pb^{2+} 结合,2/3 与 Ca^{2+} 结合;

3) 因此,过量的 Ca^{2+} 的存在会显著抑制 NTA 溶解碳酸铅中的铅。

3.1.16 Polyphosphates and Phosphonates in Water
水中的多磷酸盐和磷酸盐

重要名词和术语解析

Phosphorus: Phosphorus occurs as many oxoanions, anionic forms in combination with oxygen.

Vitreous sodium phosphate: Vitreous sodium phosphates are mixtures consisting of linear phosphate chains with from 4 to approximately 18 phosphorus atoms each.

磷:磷与氧结合,形成许多含氧阴离子、阴离子形式。

玻璃酸钠:玻璃酸钠是由线性磷酸酯链组成的混合物,每个磷酸链具有 4~18 个磷原子。

3.1.17 Complexation by Humic Substances
腐殖质的络合

知识框架

(1) Humic substances are biodegradation-resistant residues remaining from the biodegradation of plant biomass.

High-molecular mass polyelectrolytic macromolecules.

(2) When a sediment or soil is treated with base solution,

1) Humin material is humic substance not extracted.

2) Humic acid precipitates from acidified extract.

3) Fulvic acid remains in acidified solution.

(3) Humic substances are important in some natural waters.

1) Chelate metals.

2) Cause problems with metals (iron) removal from water.

3) Water color, especially when bound to iron (Gelbstoffe).

4) Insoluble humin and humic acid remove metals from solution.

5) Precursors to trihalomethane ($CHCl_3$) production in water chlorination.

(1) 腐殖质是植物生物质生物降解过程中留下的抗生物降解残留物。

高分子聚合电解质大分子。

(2) 用碱液处理沉淀物或土壤时：

1) 胡敏素是未提取的腐殖质；

2) 酸化提取物中的腐殖酸沉淀；

3) 富里酸残留在酸化溶液中。

(3) 腐殖质在某些天然水中很重要。

1) 螯合金属。

2) 导致水中金属（铁）去除问题。

3) 水的色度，尤其是与铁结合时（Gelbstoffe）。

4) 不溶性胡敏素和腐殖酸从溶液中去除金属。

5) 在水的氯化过程中产生三卤甲烷（$CHCl_3$）的前体。

重要名词和术语解析

Humic substance: The most important class of complexing agents that occur naturally are the **humic substances**. If a material containing humic substances is extracted with strong base, and the resulting solution is acidified, the products are: (1) a nonextractable plant residue called **humin**, (2) a material that precipitates from the acidified extract, called **humic acid**, and (3) an organic material that remains in the acidified solution, called **fulvic acid**.

Gelbstoffe: Yellow fulvic acid-type compounds called **Gelbstoffe** and frequently encountered along with soluble iron, are associated with color in water.

Trihalomethane: Special attention has been given to humic substances since about 1970, following the discovery of **trihalomethanes** (THMs, such as chloroform and dibromochloromethane) in water supplies.

腐殖质：自然产生的最重要的络合剂是**腐殖质**。如果用强碱提取含有腐殖质的物质，并将所得溶液酸化，则产品为：(1) 一种不可提取的植物残渣，称为**胡敏素**；(2) 从酸化提取物中沉淀出来的物质，称为**腐殖酸**；(3) 残留在酸化溶液中的有机物质，称为**富里酸**。

黄色物质：黄色富里酸型化合物称为**黄色物质**，经常与可溶性铁一起出现，与水中的颜色有关。

三卤甲烷：自从1970年左右开始在供水中发现**三卤甲烷**（THM，如氯仿和二溴氯甲烷）之后，就特别注意了腐殖质。

3.1.18 Complexation and Redox Processes 络合和氧化还原过程

知识框架

Complexation and chelation shift oxidation-reduction equilibria.

(1) Usually by stabilization of oxidized form of metal.

(2) Dissolve protective oxide coatings.

络合和螯合转移氧化还原平衡。

(1) 通常通过稳定金属的氧化形式。

(2) 溶解保护性氧化物涂层。

3.2 Oxidation/Reduction in Aquatic Chemistry
水化学中的氧化/还原

3.2.1 The Significance of Oxidation/Reduction
氧化/还原的意义

知识框架

(1) Oxidation/reduction (redox) reactions involve changes in oxidation states of elements, most easily visualized as a transfer of electrons.

(2) The reaction $Fe+Cd^{2+} \rightarrow Fe^{2+}+Cd$, can be obtained by adding the two following half-reactions:
1) Oxidation: $Fe \rightarrow Fe^{2+}+2e^-$.
2) Reduction: $Cd^{2+}+2e^- \rightarrow Cd$.
3) Overall: $Fe+Cd^{2+} \rightarrow Fe^{2+}+Cd$.

(3) In the above:
1) Fe is oxidized by losing 2 electrons to yield Fe^{2+}.
2) Cd^{2+} ion is reduced by gaining $2e^-$ to give elemental.

Oxidation/reduction reactions are significant in water

(1) Bacterially mediated oxidation of biomass depletes oxygen:
$$\{CH_2O\} + O_2 \longrightarrow CO_2 + H_2O$$

(2) Bacterially mediated reduction of solid iron oxides and hydroxides puts soluble iron in water:
$$Fe(OH)_3(s) + 3H^+ + e^- \longrightarrow Fe^{2+}(aq) + 3H_2O$$

(3) Microbially mediated oxidation of

(1) 氧化/还原（redox）反应涉及元素氧化态的变化，最容易被视为电子转移。

(2) 反应 $Fe+Cd^{2+} \rightarrow Fe^{2+}+Cd$，可以通过添加以下两个半反应来获得：
1) 氧化：$Fe \rightarrow Fe^{2+}+2e^-$。
2) 还原：$Cd^{2+}+2e^- \rightarrow Cd$。
3) 总反应：$Fe+Cd^{2+} \rightarrow Fe^{2+}+Cd$。

(3) 在上述反应中：
1) Fe 通过失去 2 个电子而被氧化，生成 Fe^{2+}。
2) Cd 通过增加 $2e^-$ 还原 Cd^{2+} 离子，得到元素 Cd。

水中的氧化/还原反应非常重要

(1) 细菌介导的生物质氧化消耗氧气：

(2) 细菌介导的固体氧化铁和氢氧化物的还原将可溶性铁释放入水中：

(3) 微生物对铵态氮的氧

ammonium nitrogen produces nitrate, which can be assimilated by algae:

$$NH_4^+ + 2O_2 \longrightarrow NO_3^- + 2H^+ + H_2O$$

(4) Two important points regarding oxidation/reduction in water:

1) Most such reactions mediated by microorganisms.

2) Close relationship with acid-base, analogies between e^- and H^+.

化会产生硝酸盐，硝酸盐可以被藻类吸收：

（4）关于水中氧化/还原的两个要点：

1）大多数此类反应是由微生物介导的。

2）与酸碱关系密切，e^-和H^+之间相似。

重要名词和术语解析

Oxidation-reduction: Oxidation-reduction (**redox**) reactions are those involving changes of oxidation states of reactants.

氧化还原：氧化还原（**redox**）反应是涉及反应物氧化态变化的反应。

3.2.2 The Electron and Redox Reactions 电子和氧化还原反应

知识框架

(1) The reaction $2Fe^{3+} + H_2 \rightleftharpoons 2Fe^{2+} + 2H^+$, can be carried out as two half-reactions in an electrochemical cell.

(2) For the electrochemical cell:

1) The left electrode is the standard hydrogen electrode, SHE.

2) Half-reaction:

$$2H^+ + 2e^- \rightleftharpoons H_2, \ E^0 = 0.00V$$

3) When all reaction participants are at unit activity (essentially [H^+] = 1.00mol/L, pressure of H_2 = 1.00atm), the potential of the SHE (E^0) is assigned a value of 0.00V.

4) The potential of the right electrode vs. SHE is the electrode potential, E.

（1）反应 $2Fe^{3+} + H_2 \rightleftharpoons 2Fe^{2+} + 2H^+$，可以在电化学电池中作为两个半反应进行。

（2）对于所示的电化学电池：

1）左电极是标准氢电极，SHE；

2）半反应：

3）当所有反应参与者均处于单位活动状态（基本上 [H^+] = 1.00mol/L，H_2 的压力等于 1.00atm）时，SHE 的电势（E^0）被指定为 0.00V；

4）右电极的电势与 SHE 的关系为电极电势 E；

5) When Fe^{3+} and Fe^{2+} are at unit activity, E is the standard electrode potential, E^0.

6) $Fe^{3+} + e^- \rightleftharpoons Fe^{2+}$, $E^0 = 0.77V$.

5) 当 Fe^{3+} 和 Fe^{2+} 处于单位活度时，E 为标准电极电位，E^0；

6) $Fe^{3+} + e^- \rightleftharpoons Fe^{2+}$，$E^0 = 0.77V$。

重要名词和术语解析

Standard hydrogen electrode: The standard electrode against which all other electrode potentials are compared. It is called the **standard hydrogen electrode**, **SHE**.

Electrode potential: The measured potential of electrode versus the standard hydrogen electrode is called the **electrode potential**, E.

Standard electrode potential: If the ions in solution are both at unit activity, the potential is the **standard electrode potential** (according to IUPAC convention, the **standard reduction potential**), E^0.

标准氢电极：与所有其他电极电势进行比较的标准电极，它被称为**标准氢电极，SHE**。

电极电势：电极相对于标准氢电极的测量电势称为**电极电势** E。

标准电极电势：如果溶液中的离子都处于单位活度，则该电势为**标准电极电势**（根据 IUPAC 约定，为标准还原电势）E^0。

3.2.3 Electron Activity and pE
电子活度和 pE

知识框架

Conceptually the negative log of electron activity, pE expresses this activity over many orders of magnitude:

从概念上讲，电子活度为负对数，pE 可在许多数量级上表示这种活度：

$$pE = -\lg(a_{e^-})$$

$$pE = \frac{E}{\dfrac{2.303RT}{F}} = \frac{E}{0.0591} \text{ (at 25℃)}$$

$$pE^0 = \frac{E^0}{\dfrac{2.303RT}{F}} = \frac{E^0}{0.0591} \text{ (at 25℃)}$$

$$2H^+(aq) + 2e^- \rightleftharpoons H_2(g), \ E^0 = 0.00V, \ pE^0 = 0.00$$

3.2.4 The Nernst Equation
能斯特方程

知识框架

$$Fe^{3+} + e^- \rightleftharpoons Fe^{2+}, \quad E^0 = 0.77V, \quad pE^0 = 13.2$$

For the $[Fe^{3+}]/[Fe^{2+}]$ electrode, which expresses pE as a function of species concentration, is:

对于 $[Fe^{3+}]/[Fe^{2+}]$ 电极，其将 pE 表示为物种浓度的函数：

$$pE = pE^0 + \frac{1}{n}\lg\frac{[Fe^{3+}]}{[Fe^{2+}]} \quad (\text{in this case } r=1) \quad (\text{在 } n=1 \text{ 的情况下})$$

$[Fe^{3+}] = 2.35 \times 10^{-3} M$, $[Fe^{2+}] = 7.85 \times 10^{-5} M$,

$$pE = 13.2 + \lg\frac{2.35 \times 10^{-3}}{7.85 \times 10^{-5}} = 14.7$$

重要名词和术语解析

<u>Nernst equation</u>: The Nernst equation is used to account for the effect of different activities upon electrode potential.

<u>能斯特方程</u>：能斯特方程用于说明不同活动对电极电势的影响。

3.2.5 Reaction Tendency: Whole Reaction from Half-Reactions
反应倾向：半反应引起的整个反应

知识框架

(1) For the overall electrode reaction in the electrochemical cell:

（1）对于电化学电池中的总电极反应：

$$2(Fe^{3+} + e^- \rightleftharpoons Fe^{2+}), \quad E^0 = 0.77V$$
$$-(2H^+ + 2e^- \rightleftharpoons H_2), \quad E^0 = 0.00V$$
$$2Fe^{3+} + H_2 \rightleftharpoons 2Fe^{2+} + 2H^+, \quad E^0 = 0.77V$$

(2) The positive value of E^0 for the overall reaction indicates that it goes to the right. H_2 reduces Fe^{3+} to Fe^{2+}.

（2）整个反应的 E^0 为正值表示它向右移动。H_2 将 Fe^{3+} 还原为 Fe^{2+}。

3.2.6 The Nernst Equation and Chemical Equilibrium
能斯特方程和化学平衡

知识框架

Consider the system illustrated in Figure 3.1. 　　考虑图 3.1 所示的系统。

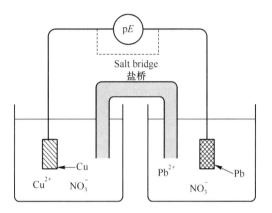

Figure 3.1　Schematic diagram of a REDOX reaction system

图 3.1　氧化还原反应系统示意图

$$Cu^{2+}+2e^- \rightleftharpoons Cu,\ pE^0=5.71$$
$$-(Pb^{2+}+2e^- \rightleftharpoons Pb,\ pE^0=-2.13)$$
$$Cu^{2+}+Pb \rightleftharpoons Cu+Pb^{2+},\ pE^0=7.84$$

The equilibrium constant for this reaction is: 　　该反应的平衡常数为：

$$K=\frac{[Pb^{2+}]}{[Cu^{2+}]}$$

$$pE=pE^0+\frac{1}{n}\lg\frac{[Cu^{2+}]}{[Pb^{2+}]}=7.84+\frac{1}{2}\lg\frac{[Cu^{2+}]}{[Pb^{2+}]}$$

When $pE=0$, the system is at equilibrium: 　　当 $pE=0$ 时，系统处于平衡状态：

$$pE=0.00=7.84-\frac{1}{2}\lg\frac{[Pb^{2+}]}{[Cu^{2+}]}=7.84-\frac{1}{2}\lg K,\ \lg K=15.7$$

In general, $\lg K=npE^0$.　　一般来说，$\lg K=npE^0$。

3.2.7 The Relationship of pE to Free Energy
pE 与自由能的关系

知识框架

$$DG = -2.303nRT(pE)$$
$$DG^0 = -2.303nRT(pE^0)$$

3.2.8 Reactions in Terms of One Electron-Mole
一个电子摩尔的反应

知识框架

(1) Writing oxidation/reduction reactions on the basis of one electron-mole enables their comparison on a common basis.

Especially significant for free energy comparisons.

(2) Consider the reaction：
$$NH_4^+ + 2O_2 \rightleftharpoons NO_3^- + 2H^+ + H_2O, \quad pE^0 = 5.85$$

(3) The N atom changes oxidation state from -3 to $+5$ corresponding to a transfer of 8 electrons, so the reaction in terms of 1 mole of electrons is：
$$\frac{1}{8}NH_4^+ + \frac{1}{4}O_2 \rightleftharpoons \frac{1}{8}NO_3^- + \frac{1}{4}H^+ + \frac{1}{8}H_2O, \quad pE^0 = 5.85$$

(4) For a reaction written as one electron-mole ($n=1$)：
1) $DG^0 = -2.303RT(pE^0)$.
2) $\lg K = pE^0$.

(1) 在一个电子摩尔的基础上写氧化/还原反应，可以在一个通用的基础上进行比较。

对于自由能比较特别重要。

(2) 考虑反应：

(3) N 原子将氧化态从 -3 价改变为 $+5$ 价，对应 8 个电子的转移，因此以 1mol 电子表示的反应为：

(4) 写为一个电子摩尔 ($n=1$) 的反应：
1) $DG^0 = -2.303RT(pE^0)$；
2) $\lg K = pE^0$。

重要名词和术语解析

E_H **value**: It is true that one may place a platinum electrode and a reference electrode in water and measure a potential. This potential, referred to the standard hydrogen electrode, is the so-called E_H **value**.

E_H **值**：可以将铂电极和参比电极放在水中并测量电位。这个称为标准氢电极的电势就是所谓的 E_H 值。

3.2.9 The Limits of pE in Water
水中 pE 的限值

知识框架

(1) Oxidizing limit of H_2O stability:

$$2H_2O \rightleftharpoons O_2 + 4H^+ + 4e^-$$

(2) Reducing limit of H_2O stability

$$2H_2O + 2e^- \rightleftharpoons H_2 + 2OH^-$$

(3) For the oxidizing limit set pressure of O_2 as 1.00 atm[❶]:

$$\frac{1}{4}O_2 + H^+ + e^- \rightleftharpoons \frac{1}{2}H_2O, \quad pE^0 = 20.75$$

$$pE = pE^0 + \lg(P_{O_2}^{\frac{1}{4}}[H^+])$$

Plot pE = 20.75 − pH.

(4) For the reducing limit set pressure of H_2 as 1.00 atm:

$$H^+ + e^- \rightleftharpoons \frac{1}{2}H_2, \quad pE^0 = 0.00$$

$$pE = pE^0 + \lg[H^+]$$

Plot pE = −pH.

(1) H_2O 稳定的氧化极限:

(2) H_2O 稳定的还原极限:

(3) 对于氧化极限, 将 O_2 的压力设置为 1.00 atm[❶]:

作图 pE = 20.75 − pH

(4) 对于还原极限, 将 H_2 的压力设置为 1.00 atm:

作图 pE = −pH。

3.2.10 pE Value in Natural Water Systems
天然水体系中的 pE 值

知识框架

$$pE = pE^0 + \lg(P_{O_2}^{\frac{1}{4}}[H^+])$$

(1) Consider pH = 7.00 water in equilibrium with atmospheric oxygen (0.21 atm partial pressure)

$$\frac{1}{4}O_2 + H^+ + e^- \rightleftharpoons \frac{1}{2}H_2O, \quad pE^0 = 20.75$$

$$pE = pE^0 + \lg(P_{O_2}^{\frac{1}{4}}[H^+])$$

(1) 考虑 pH = 7.00 的水中, 氧分压与大气中的氧分压保持平衡 (分压为 0.21 atm)

---

❶ 1atm = 101.325kPa。

(2) Substituting 0.21 for O_2 pressure and $[H^+] = 1.00 \times 10^{-7} M$.

$pE = 13.8$.

(3) Consider anoxic water at $pH = 7.00$ in which the pressures of CH_4 and CO_2 are equal:

$$pE = 2.87 + \lg \frac{P_{CO_2}^{\frac{1}{8}}[H^+]}{P_{CH_4}^{\frac{1}{8}}} = 2.87 + \lg[H^+] = 2.87 - 7.00 = -4.13$$

（2）用 0.21 代替 O_2 分压，且 $[H^+] = 1.00 \times 10^{-7} M$。

$pE = 13.8$。

（3）考虑 $pH = 7.00$ 的缺氧水，其中 CH_4 和 CO_2 的压力相等：

3.2.11 pE-pH Diagrams
pE-pH 图

知识框架

Discussion of Iron pE-pH Diagram

(1) Dashed lines are thermodynamic limits of water stability.

(2) Fe^{3+} is stable in a small region at high pE and low pH (such as acid mine water in contact with the atmosphere).

(3) Fe^{2+} ion is stable over a relatively large range of pH and relatively low pE often encountered in groundwater and anoxic water in the bottom of bodies of water; bringing this water into contact with the atmosphere results in precipitation of $Fe(OH)_3$.

(4) Solid $Fe(OH)_3$ predominates over a very large region reflecting the very low solubilities of iron(Ⅲ) hydroxides and oxides.

(5) Solid $Fe(OH)_2$ has a limited region of stability; in most aquatic systems FeS or $Fe(CO_3)$ are the iron(Ⅱ) compounds found in this region.

铁 pE-pH 图的讨论

（1）虚线是水稳定的热力学极限。

（2）Fe^{3+} 在高 pE 值和低 pH 值的小范围内稳定（如与大气接触的酸性矿井水）。

（3）Fe^{2+} 离子在相对较大的 pH 值范围和相对较低的 pE 值（在水体底部的地下水和缺氧水中经常遇到）稳定；这种水与大气接触，会导致 $Fe(OH)_3$ 沉淀。

（4）固态 $Fe(OH)_3$ 在很大的区域占主导地位，反映出三价铁氢氧化物和氧化物的溶解度非常低。

（5）固态 $Fe(OH)_2$ 的稳定性有限；在大多数水生系统中，FeS 或 $Fe(CO_3)$ 是该区域发现的二价铁化合物。

3.2.12 Humic Substances as Natural Reductants
腐殖质作为天然还原剂

知识框架

Humic substances may function as reducing agents in natural water systems.

May be due to the quinone/hydroquinone group.

腐殖质可能在天然水系统中充当还原剂。

可能归因于醌/氢醌基团。

3.2.13 Photochemical Processes in Oxidation/Reduction
氧化/还原中的光化学过程

知识框架

Superoxide ion, $O_2^-\cdot$.

(1) Generated by photochemically excited humic matter acting upon dissolved O_2.

(2) Reacts in water to produce.

(3) H_2O_2 undergoes the Fenton reaction in water to produce reactive hydroxyl radical, $HO\cdot$.

(4) $Fe(II) + H_2O_2 \rightarrow Fe(III) + OH^- + HO\cdot$.

超氧离子，$O_2^-\cdot$。

(1) 由光化学激发的腐殖质作用于溶解氧产生。

(2) H_2O_2 与水反应生成 H_2O_2。

(3) H_2O_2 在水中进行 Fenton 反应，生成反应性羟基自由基 $HO\cdot$。

(4) $Fe(II) + H_2O_2 \rightarrow Fe(III) + OH^- + HO\cdot$。

3.2.14 Corrosion
腐蚀

知识框架

Corrosion takes place when an electrochemical cell is set up on the surface of a metal:

(1) Metal oxidized at the anode:

$$M \longrightarrow M^{2+} + 2e^-$$

(2) Several possible cathodic reactions often involving oxygen species:

M 当将电化学电池安装在金属 M 上时会发生腐蚀：

(1) 阳极氧化的金属：

(2) 几种涉及氧气的可能的阴极反应有：

$$O_2 + 2H_2O + 4e^- \longrightarrow 4OH^-$$
$$O_2 + 4H^+ + 4e^- \longrightarrow 2H_2O$$
$$O_2 + 2H_2O + 2e^- \longrightarrow 2OH^- + H_2O_2$$

(3) Oxygen may accelerate corrosion by participating in cathodic reactions.

(4) In some cases oxygen inhibits corrosion by forming protective oxide coatings.

(5) Bacteria are commonly involved in corrosion and set up electrochemical cells in which corrosion occurs.

(3) 氧气可能通过参与阴极反应而加速腐蚀。

(4) 在某些情况下, 氧气会通过形成保护性氧化物涂层来抑制腐蚀。

(5) 细菌通常参与腐蚀并组成发生腐蚀的电化学电池。

重要名词和术语解析

Corrosion: One of the most damaging redox phenomena is **corrosion**, defined as the destructive alteration of metal through interactions with its surroundings.

腐蚀: 最具破坏性的氧化还原现象之一是**腐蚀**, 定义为金属通过与周围环境的相互作用而发生的破坏性变化。

3.3 Phase Interactions in Aquatic Chemistry
水化学中的相间作用

3.3.1 Importance and Formation of Sediments
沉积物的重要性和形成

知识框架

Sediments are:

(1) Layers of relatively finely divided matter.

(2) Cover bottoms of various bodies of water.

(3) Generally mixtures of clay, silt, sand, organic matter.

(4) Various organisms.

(5) Pollutants including heavy metals, organics.

沉积物是:

(1) 相对细分的物质层;

(2) 盖住各种水体的底部;

(3) 通常是黏土、淤泥、沙子、有机物的混合物;

(4) 各种生物;

(5) 污染物, 包括重金属、有机物;

(6) Transfer to organisms directly or through pore water.

Formation of sediments

(1) Physical transfer of material.

(2) Chemical precipitation.

(3) Biochemical processes such as photosynthesis producing biomass and solid $CaCO_3$, action of anoxic bacteria producing solid FeS.

Organic and Carbonaceous Sedimentary Materials

(1) Particularly important for binding organic pollutants.

(2) Organics may be held for many years.

(3) Black carbon from combustion.

(6) 直接或通过孔隙水迁移至生物。

沉积物的形成

(1) 材料的物理迁移。

(2) 化学沉淀。

(3) 生化过程，例如光合作用产生生物质和固体 $CaCO_3$，缺氧细菌产生固体 FeS 的作用。

有机和碳质沉积物

(1) 对于结合有机污染物特别重要。

(2) 有机物可能会保存很多年。

(3) 燃烧产生的黑炭。

重要名词和术语解析

Sediment: Sediments are the layers of relatively finely divided matter covering the bottoms of rivers, streams, lakes, reservoirs, bays, estuaries, and oceans.

Pore water: The portion of substances held in sediments that is probably most available to organisms is that contained in **pore water**, contained in microscopic pores within the sediment mass.

沉积物：沉积物是指相对细碎的物质层，覆盖了河流、溪流、湖泊、水库、海湾、河口和海洋的底部。

孔隙水：沉积物中所含物质最可能为生物体所用的部分是孔隙水中所含的物质，这些物质包含在沉积物团块的微小孔隙中。

3.3.2 Solubilities
溶解度

知识框架

(1) Solubilities of solids：

(1) 固体溶解度：

1) From solubility products.
2) Intrinsic solubility, example of $CaSO_4$:
$$S = [Ca^{2+}] + [CaSO_4]$$
Where, $[Ca^{2+}]$ from solubility product, $[CaSO_4]$ from intrinsic solubility.

(2) Solubilities of ionic solids affected by several factors, example of $PbCO_3$:
1) Increased by chelation of metal:
$$Pb^{2+} + T^{3-} \rightleftharpoons PbT^-$$
2) Increased by reaction of anion:
$$PbCO_3 + H^+ \rightleftharpoons Pb^{2+} + HCO_3^-$$
3) Presence of common ion:
$$HCO_3^- \rightleftharpoons H^+ + CO_3^{2-}$$

1) 来自溶度积。
2) 固有溶解度, 例如 $CaSO_4$:

式中, $[Ca^{2+}]$ 来自溶度积; $[CaSO_4]$ 来自固有溶解度。

(2) 离子固体的溶解度受多种因素影响, 如 $PbCO_3$:
1) 因金属螯合而增加:

2) 因阴离子反应而增加:

3) 常见离子的存在:

Solubilities of Gases

Henry's law: At constant temperature the solubility of a gas in a liquid is proportional to the partial pressure of the gas in contact with the liquid.

(1) $X(g) \leftrightarrow X(aq)$.
(2) $[X(aq)] = KP_X$.

Increased by acid-base reactions:
$$NH_3(g) + H^+ \rightleftharpoons NH_4^+(aq)$$
$$SO_2(g) + HCO_3^- \text{ (alkalinity 碱度)} \rightleftharpoons HSO_3^-(aq) + CO_2$$

气体溶解度

亨利定律: 在恒定温度下, 气体在液体中的溶解度和与液体接触的气体的分压成正比。

(1) $X(g) \leftrightarrow X(aq)$。
(2) $[X(aq)] = KP_X$。

酸碱反应增加溶解度:

Concentration of O_2 in water in contact with air at 25℃

25℃下与空气接触的水中 O_2 的浓度

Partial pressure of water at 25℃
25℃下水的分压

Mole fraction of O_2 in the atmosphere
大气中O_2的摩尔分数

$$P_{O_2} = (1.0000\text{atm}❶ - 0.0313\text{atm}) \times 0.2095 = 0.2029\text{atm}$$
$$[O_2(aq)] = KP_{O_2} = 1.28 \times 10^{-3} \text{mol}/(L \cdot \text{atm})^{-1} \times 0.2029\text{atm}$$
$$[O_2(aq)] = 2.60 \times 10^{-4} \text{mol/L}$$

---

❶ 1atm = 101.325kPa。

(Concentration in mg/L)　mg O_2 in a mole of O_2
(浓度用mg/L表示)　1mol O_2中的毫克数

$2.60×10^{-4}$ mol/L $×3.2×10^4$ +mg/mol=8.32mg/L

The Clausius-Clapeyron equation for gas solubilities C_1 and C_2 at absolute temperatures of T_1 and T_2 where R is the gas constant and DH is the heat of solution:

绝对温度 T_1 和 T_2 时，气体溶解度 C_1 和 C_2 的 Clausius-Clapeyron 方程，其中 R 为气体常数，DH 为溶液热，即：

$$\lg \frac{C_2}{C_1} = \frac{\Delta H}{2.303R}\left[\frac{1}{T_1} - \frac{1}{T_2}\right]$$

重要名词和术语解析

Solubility product: An equilibrium constant in this form that expresses the solubility of a solid that forms ions in water is a **solubility product** and is designated K_{sp}.

Intrinsic solubility: Intrinsic solubilities account for the fact that a significant portion of the solubility of an ionic solid is due to the dissolution of the neutral form of the salt and must be added to the solubility calculated from K_{sp} to obtain the total solubility.

溶度积：表示在水中形成离子的固体的溶解度的平衡常数是**溶度积**，记作 K_{sp}。

固有溶解度：**固有溶解度**说明以下事实：离子固体的大部分溶解度是由于盐的中性形式的溶解所致，必须将其添加到由 K_{sp} 计算得到的溶解度中，以获得总溶解度。

3.3.3 Colloidal Particles in Water
水中的胶体颗粒

知识框架

Size range of 0.001~1 micrometers.

尺寸为 0.001~1 μm。

Include
(1) Minerals.
(2) Microorganisms.
(3) Organic matter.
(4) Proteinaceous material.

包括
(1) 矿物质。
(2) 微生物。
(3) 有机物。
(4) 蛋白质物质。

Important characteristics

(1) Light scattering (Tyndall effect).

(2) High area.

(3) High interfacial area.

(4) High surface/charge density ratio.

Important, widespread in natural waters and wastewaters.

Colloid-facilitated transport of pollutants.

Colloids are widespread in water and wastewater.

Kinds of colloids

(1) Hydrophobic.

(2) Hydrophilic.

(3) Association.

Colloid stability

Stabilized by attraction to water and by surface charge.

Colloidal particles acquire charge by:

(1) Surface chemical reaction, often involving H^+.

(2) Ion absorption.

(3) Ion replacement [such as Al(Ⅲ) for Si(Ⅳ)] in clays.

重要特征

(1) 光散射（丁达尔效应）。

(2) 高面积。

(3) 高界面面积。

(4) 高表面/电荷密度比。

广泛存在于天然水和废水中。

胶体促进污染物的迁移。

胶体广泛存在于水和废水中。

胶体种类

(1) 疏水。

(2) 亲水。

(3) 缔合。

胶体稳定性

通过吸引水和表面电荷来稳定。

胶体粒子获得电荷通过：

(1) 表面化学反应，通常涉及 H^+。

(2) 离子吸收。

(3) 黏土中的离子置换[如 Al(Ⅲ) 替代 Si(Ⅳ)]。

重要名词和术语解析

Colloidal particle: Such particles, which have some characteristics of both species in solution and larger particles in suspension, which range in diameter from about 0.001μm to about 1μm, and which scatter white light as a light blue hue observed at right angles to the incident light, are classified as **colloidal particles**.

胶体粒子：这种粒子具有溶液中物质和悬浮液中较大的粒子两种物质的一些特性，直径在 0.001μm 到 1μm 之间，与入射光成合适的角度观察时，将白色光散射成淡蓝色调的粒子，被归类为**胶体粒子**。

Tyndall effect: The characteristic light-scattering phenomenon of colloids results from their being the same order of size as the wavelength of light and is called the **Tyndall effect**.

Hydrophilic colloid: Hydrophilic colloids generally consist of macromolecules, such as proteins and synthetic polymers, that are characterized by strong interaction with water resulting in spontaneous formation of colloids when they are placed in water.

Hydrophobic colloid: Hydrophobic colloids interact to a lesser extent with water and are stable because of their positive or negative electrical charges. The charged surface of the colloidal particle and the counter-ions that surround it compose an electrical double layer, which causes the particles to repel each other.

Association colloid: Association colloids consist of special aggregates of ions and molecules called micelles.

Critical micelle concentration: micelles form when a certain concentration of surfactant species, typically around 1×10^{-3}, is reached. The concentration at which this occurs is called the **critical micelle concentration**.

Hydration and surface charge: the two main phenomena contributing to the stabilization of colloids are **hydration** and **surface charge**.

Chemical reaction at the particle surface: One of the three major ways in which a particle may acquire a surface charge is by **chemical reaction at the particle surface**.

丁达尔效应：胶体的特征性光散射现象是由于胶体的大小与光的波长相同，因此称为**丁达尔效应**。

亲水性胶体：**亲水性胶体**通常由大分子组成，如蛋白质和合成聚合物，其特征是与水的相互作用强，将其置于水中时会自发形成胶体。

疏水性胶体：**疏水性胶体**与水的相互作用程度较小，并且由于其带正电荷或负电荷而稳定。胶体颗粒的带电表面和围绕它的抗衡离子组成了一个双电层，使颗粒彼此排斥。

缔合胶体：**缔合胶体**由离子和称为胶束的分子的特殊聚集体组成。

临界胶束浓度：当达到一定浓度的表面活性剂物质（通常约为 1×10^{-3}）时，就会形成胶束。发生这种情况的浓度称为**临界胶束浓度**。

水化和表面电荷：促进胶体稳定的两个主要现象是**水合**和**表面电荷**。

颗粒表面的化学反应：颗粒获得表面电荷的三种主要方法之一是通过**颗粒表面的化学反应**。

Zero point of charge: At some intermediate pH value, called the **zero point of charge (ZPC)**, colloidal particles of a given hydroxide will have a net charge of zero, which favors aggregation of particles and precipitation of a bulk solid.

Ion absorption: **Ion absorption** is a second way in which colloidal particles become charged.

Ion replacement: **Ion replacement** is a third way in which a colloidal particle may gain a net charge.

零电荷点：在某个中等的，称为**零电荷点（ZPC）**的 pH 下，给定氢氧化物的胶体颗粒的净电荷为零，这有利于颗粒的聚集和大块固体的沉淀。

离子吸收：**离子吸收**是胶体粒子带电的第二种方式。

离子置换：**离子置换**是胶体粒子获得净电荷的第三种方式。

3.3.4 The Colloidal Properties of Clays 黏土的胶体性质

知识框架

(1) Clays are widespread as colloidal particles in water and as solids in sediments.

1) Secondary minerals.
2) Hydrated aluminum and silicon oxides.

(2) Common clays include:
1) Kaolinite: $Al_2(OH)_4Si_2O_5$.
2) Montmorillonite: $Al_2(OH)_2Si_4O_{10}$.
3) Nontronite: $Fe_2(OH)_2Si_4O_{10}$.
4) Hydrous mica: $KAl_2(OH)_2(AlSi_3)O_{10}$.

(3) Unit layers in clay structures.

(4) Clays acquire charge usually by substitution of Al(Ⅲ) for Si(Ⅳ):
1) Exchangeable cations, such as H^+, K^+, and NH_4^+.
2) Cation exchange capacity.

(1) 黏土作为水中的胶体颗粒和沉积物中的固体而广泛分布。

1) 次生矿物。
2) 水合铝和硅氧化物。

(2) 常见的黏土包括：
1) 高岭石：$Al_2(OH)_4Si_2O_5$；
2) 蒙脱石：$Al_2(OH)_2Si_4O_{10}$；
3) 绿脱石：$Fe_2(OH)_2Si_4O_{10}$；
4) 水合云母：$KAl_2(OH)_2(AlSi_3)O_{10}$。

(3) 黏土结构中的单元层。

(4) 黏土通常通过用 Al(Ⅲ) 代替 Si(Ⅳ) 获得电荷
1) 可交换阳离子，例如 H^+、K^+ 和 NH_4^+；
2) 阳离子交换量。

重要名词和术语解析

Clay: **Clays** consist largely of hydrated aluminum and silicon oxides and are **secondary minerals**, which are formed by weathering and other processes acting on primary rocks.

Unit layer: Units of two or three sheets make up **unit layers**.

Exchangeable cation: **Exchangeable cations** are exchangeable for other cations in water.

Cation-exchange capacity: The amount of exchangeable cations, expressed as milliequivalents (of monovalent cations) per 100 g of dry clay, is called the **cation-exchange capacity**, **CEC**, of the clay and is a very important characteristic of colloids and sediments that have cation-exchange capabilities.

黏土：黏土主要由水合铝和硅氧化物组成，是**次生矿物**，是通过风化和其他作用于原始岩石上的过程形成的。

单位层：两到三片层组成**单位层**。

可交换阳离子：可交换阳离子可与水中的其他阳离子交换。

阳离子交换量：可交换阳离子的量，以每100g干黏土的毫当量（一价阳离子）表示，称为黏土的**阳离子交换量CEC**，具有阳离子交换能力是胶体和沉积物的非常重要的特征。

3.3.5 Aggregation of Particles
粒子聚集

知识框架

Important in water

(1) Example 1: Settling of waste biomass in wastewater treatment.

(2) Example 2: Formation of sediments from river water entering oceans.

Mechanisms of aggregation

(1) Coagulation from reduction of surface charge repulsion.

(2) Flocculation with bridging compounds that produce floc networks.

在水中重要

(1) 例1：废水处理中废物生物质的沉降。

(2) 例2：河水进入海洋形成沉积物。

聚集机制

(1) 减少表面电荷排斥引起的凝聚。

(2) 与产生絮凝网络的桥联化合物絮凝。

(3) Flocculation is facilitated by synthetic and natural polyelectrolytes.

(3) 合成和天然聚电解质可促进絮凝。

重要名词和术语解析

Coagulation: Coagulation involves the reduction of this electrostatic repulsion such that colloidal particles of identical materials may aggregate.

Flocculation: Flocculation uses **bridging compounds**, which form chemically bonded links between colloidal particles and enmesh the particles in relatively large masses called **floc networks**.

Double-layer compression: Because of the double layer of electrical charge surrounding a charged particle, this aggregation mechanism is sometimes called **double-layer compression**.

Polyelectrolyte: Polyelectrolytes of both natural and synthetic origin may cause colloids to flocculate.

Bacterial floc: Bacterial floc consists of aggregated bacterial cells that have settled from the water.

凝聚：凝聚涉及静电排斥的减少，使得相同材料的胶体颗粒可能会聚集。

絮凝：絮凝使用桥接化合物，它们在胶体颗粒之间形成化学键连接，并将相对较大的质量的颗粒以称为**絮凝网络**的形式啮合。

压缩双电层：由于带电粒子周围的电荷是双层的，因此这种聚集机制有时称为**压缩双电层**。

聚合电解质：天然和合成来源的**聚合电解质**都可能导致胶体絮凝。

细菌絮凝物：细菌絮凝物由从水中沉降的聚集细菌细胞组成。

3.3.6 Surface Sorption by Solids
固体表面吸附

知识框架

(1) Many of the effects of colloidal and sedimentary solids in contact with water have to do with their sorption of solutes.

(2) Metals are sorbed by solids, particularly metal oxides.

1) Nonspecific ion exchange adsorption.

2) Complexation with surface —OH groups.

(1) 与水接触的胶体和沉积固体的许多影响，都与溶质的吸附有关。

(2) 金属被固体吸附，特别是金属氧化物。

1) 非特异性离子交换吸附。

2) 与表面 —OH 基团络合。

3) Coprecipitation with the metal oxide.

4) Discrete oxide or hydroxide sorbed to metal.

(3) Hydrated manganese (Ⅳ) and iron (Ⅲ) oxides are good sorbents, especially when freshly precipitated.

Freshly precipitated MnO_2 may have a surface area of several hundred square meters per gram.

(4) Anions are also sorbed by solids.

Usually with less specific bonding than metals.

3.3.7 Solute Exchange with Bottom Sediments 与底部沉积物的溶质交换

知识框架

(1) Bottom sediments are important sources and sinks of inorganic and organic matter in streams, fresh-water impoundments, estuaries and oceans.

1) Generally anoxic (reducing conditions).

2) Generally high levels of organic matter.

(2) Cation exchange capacity (CEC) expresses the capacity of a sediment to sorb cations.

Expressed as milliequivalents per 100g solid.

(3) Exchangeable cation status (ECS) refers to specific ions held by sediments.

Common cations held by sediments are H^+, K^+, NH_4^+, Ca^{2+}, Mg^{2+}, Fe^{2+}, Mn^{2+}, Zn^{2+}, Cu^{2+}, Ni^{2+}.

3) 与金属氧化物共沉淀。

4) 吸附在金属上的离散氧化物或氢氧化物。

(3) 水合锰(Ⅳ)和铁(Ⅲ)氧化物是良好的吸附剂，尤其是在刚沉淀时。

刚沉淀的 MnO_2 的表面积可能为每克数百平方米。

(4) 阴离子也被固体吸附。

通常比金属具有更少的特异性结合。

(1) 底部沉积物是溪流、淡水蓄水池、河口和海洋中无机和有机物的重要的源和汇。

1) 一般缺氧（还原条件）。

2) 有机物含量普遍较高。

(2) 阳离子交换量（CEC）表示沉积物吸收阳离子的能力。

表示为每100g固体的毫当量。

(3) 可交换阳离子状态（ECS）是指沉积物所包含的特定离子。

沉积物所包含的常见阳离子为 H^+、K^+、NH_4^+、Ca^{2+}、Mg^{2+}、Fe^{2+}、Mn^{2+}、Zn^{2+}、Cu^{2+}、Ni^{2+}。

(4) Sediments act as buffers by exchanging H^+

Trace-Level Metals in Suspended Matter and Sediments

(1) Trace metals held in sediments and colloidal suspensions include cadmium, chromium, cobalt, copper, manganese, molybdenum, and nickel.

(2) Metals held in suspended particles less available than those in solution but more so than those in sediments.

(3) pE is an important factor.

1) High pE (oxic, oxidizing): Oxides, hydroxides, and carbonates such as HgO, $Cu(OH)_2 \cdot CuCO_3$.

2) Low pE (anoxic, reducing): Sulfides predominate such as CdS, PbS.

Phosphorus Exchange with Bottom Sediments

(1) Important in algal growth and eutrophication.

(2) Forms of phosphorus in sediments:

1) Phosphate minerals, $Ca_5OH(PO_4)_3$.

2) Nonoccluded phosphorus such as PO_4^{3-} held on mineral surfaces.

3) Occluded phosphorus with orthophosphate ions contained within mineral matrix, such as in aluminosilicates.

4) Organic phosphorus in biomass (usually algal or bacterial).

(4) 沉积物通过交换 H^+ 起到缓冲作用。

悬浮物和沉积物中的痕量金属

(1) 沉积物和胶体悬浮物中所含的微量金属包括镉、铬、钴、铜、锰、钼和镍。

(2) 悬浮颗粒中的金属比溶液中的金属少,但比沉积物中的金属多。

(3) pE 是重要因素。

1) 高 pE (有氧,氧化):氧化物,氢氧化物和碳酸盐,如 HgO、$Cu(OH)_2 \cdot CuCO_3$。

2) 低 pE (缺氧,还原):硫化物占主导地位,例如 CdS、PbS。

底部沉积物的磷交换

(1) 对藻类生长和富营养化很重要。

(2) 沉积物中磷的形式:

1) 磷酸盐矿物,$Ca_5OH(PO_4)_3$。

2) 吸附在矿物表面的非闭蓄态磷,例如 PO_4^{3-}。

3) 包含正磷酸盐离子的闭蓄态磷被保留在矿物基质(如硅铝酸盐)中。

4) 生物质中的有机磷(通常为藻类或细菌)。

Organic Compounds on Sediments and Suspended Matter

(1) Sediments as sinks and repositories of organic matter.

(2) Colloids may transport organic matter.

(3) Sorption affects degradation of organic matter.

(4) Common sorbents are clays, humic substances and complexes between clays and humic substances.

(5) Sorption generally proportional to water solubility.

Bound residues of organics

(1) Immobilization.

(2) Detoxification.

Bioavailability of Sediment Contaminants

(1) The facility with which a substance may be taken up by organisms.

(2) May be direct or through water.

沉积物和悬浮物上的有机化合物

(1) 沉积物作为有机物的汇和库。

(2) 胶体可能迁移有机物。

(3) 吸附会影响有机物的降解。

(4) 常见的吸附剂是黏土、腐殖质以及黏土和腐殖质之间的复合物。

(5) 吸附通常与水溶性成正比。

有机物的残留物

(1) 固定。

(2) 解毒。

沉积物污染物的生物利用度

(1) 有机体可以吸收物质的设施。

(2) 可以是直接的或通过水。

重要名词和术语解析

Cation-exchange capacity: Cation-exchange capacity (CEC) measures the capacity of a solid, such as a sediment, to sorb cations. It varies with pH and with salt concentration.

Exchangeable cation status: Exchangeable cation status (ECS), refers to the amounts of specific ions bonded to a given amount of sediment.

Phosphate minerals: particularly hydroxyapatite, $Ca_5OH(PO_4)_3$.

Nonoccluded phosphorus: Nonoccluded

阳离子交换量：阳离子交换量（CEC）衡量固体（如沉淀物）吸收阳离子的能力。它随 pH 和盐浓度而变化。

可交换阳离子状态：可交换阳离子状态（ECS）是指与给定数量的沉积物结合的特定离子的量。

磷酸盐矿物：尤其是羟基磷灰石 $Ca_5OH(PO_4)_3$。

非闭蓄态磷：结合在 SiO^+ 或

phosphorus such as orthophosphate ion bound to the surface of SiO^+ or $CaCO_3$. Such phosphorus is generally more soluble and more available than occluded phosphorus.

Occluded phosphorus: Occluded phosphorus consisting of orthophosphate ions contained within the matrix structures of amorphous hydrated oxides of iron and aluminum and amorphous aluminosilicates. Such phosphorus is not as readily available as nonoccluded phosphorus.

Organic phosphorus: Organic phosphorus incorporated within aquatic biomass, usually of algal or bacterial origin.

$CaCO_3$ 表面的**非闭蓄态磷**，如正磷酸根离子。此类磷通常比闭蓄态磷更易溶且更易获得。

闭蓄态磷：**闭蓄态磷**由铁和铝的无定形水合氧化物和无定形硅铝酸盐的基质结构中包含的正磷酸盐离子组成。这种磷不如非闭蓄态磷容易获得。

有机磷：掺入水生物质中的**有机磷**，通常来自藻类或细菌。

3.3.8　Interstitial Water
间隙水

知识框架

Water held in voids and pores in sediments

(1) Reflects chemical and biochemical conditions in sediment.

(2) Products of decomposition and mineralization of planktonic biomass.

(3) Largely through activity of anoxic bacteria in sediments.

(4) Gases in interstitial water.

(5) Usually virtually no O_2.

(6) N_2 usually stripped by action of anoxic bacteria producing CO_2 and CH_4 N_2.

沉积在空隙和孔中的水

（1）反映沉积物中的化学和生化条件。

（2）浮游生物质的分解和矿化产物。

（3）主要是通过沉积物中缺氧细菌的活动。

（4）间隙水中的气体。

（5）通常几乎没有氧气。

（6）通常通过产生 CO_2 和 CH_4 的缺氧细菌的作用而脱除。

重要名词和术语解析

Interstitial water: Interstitial water or pore water consisting of water held by sediments is

间隙水：由沉积物中的水组成的**间隙水**或孔隙水是天然水系

an important reservoir for gases in natural water systems.

统中气体的重要储层。

3.3.9 Phase Interactions in Chemical Fate and Transport
化学归宿和迁移中的相间作用

知识框架

Hydrosphere is particularly important in fate and transport:

(1) Rivers move dissolved and suspended substances long distances.

(2) Water bodies are repositories, but movement still occurs.

在归宿和迁移中，水圈特别重要：

(1) 河流将溶解和悬浮的物质长距离移动；

(2) 水体是储存库，但仍会移动。

Exchange with the Atmosphere

(1) Gases from air to water:

1) Oxygen required by fish.

2) Carbon dioxide required by algae.

3) Air pollutants including acid gases and particles.

(2) Gases from water to air:

1) O_2 from algal photosynthesis.

2) CO_2 from microbial degradation of organic matter.

3) H_2S from anoxic microbial reduction of SO_4^{2-}.

4) Volatile organic water pollutants.

与大气交换

(1) 从空气到水的气体：

1) 鱼所需的氧气；

2) 藻类需要二氧化碳；

3) 空气污染物，包括酸性气体和微粒。

(2) 从水到空气的气体：

1) 藻类光合作用产生的 O_2；

2) 微生物降解有机物产生的 CO_2；

3) 缺氧微生物还原 SO_4^{2-} 产生的 H_2S；

4) 挥发性有机水污染物。

Exchange with Sediments

Pollutants are incorporated with particles as they form and settle in water and are placed in sediments.

与沉积物交换

污染物在形成和沉淀于水中时会与颗粒结合在一起，并置于沉积物中。

3.4　Aquatic Microbial Biochemistry
水生微生物生物化学

3.4.1　Aquatic Biochemical Processes
水生生化过程

知识框架

(1) Microorganisms are living catalysts for chemical processes in water and soil.

1) Bacteria.
2) Fungi.
3) Protozoa.
4) Algae.

(2) Most major chemical processes in water are mediated by microorganisms.

1) Oxidation/reduction.
2) Reactions involving organic matter.

(3) Algae are primary producers of organic matter (biomass) in water.

(4) Microbial processes form many sediments and mineral deposits.

(5) Microbial processes are involved in biogeochemical process.

Example: cyanobacteria in early history of life on Earth produced all the atmosphere's O_2.

Characteristics of microorganisms in water

(1) Pathogens, including viruses, must be eliminated from drinking water.

(2) Microorganisms in water are prokaryotes

(1) 微生物是水和土壤中化学过程的活催化剂。

1) 细菌;
2) 真菌;
3) 原生动物;
4) 藻类。

(2) 水中大多数主要化学过程是由微生物介导的。

1) 氧化/还原。
2) 涉及有机物的反应。

(3) 藻类是水中有机物(生物质)的主要生产者。

(4) 微生物过程形成许多沉积物和矿物质沉积物。

(5) 微生物过程涉及生物地球化学过程。

例:地球生命早期的蓝细菌产生了大气中所有的 O_2。

水中微生物的特征

(1) 必须从饮用水中消除包括病毒在内的病原体。

(2) 水中的微生物是原核

(bacteria) and eukaryotes, (protozoa, fungi, algae).

(3) Algae and photosynthetic bacteria are producers meaning that they produce biomass, {CH_2O}, from inorganic compounds.

(4) Fungi and most bacteria are reducers that break down {CH_2O} to CO_2 and other simple inorganic compounds.

生物（细菌）和真核生物（原生动物、真菌、藻类）。

（3）藻类和光合细菌是生产者，它们从无机化合物生产生物质{CH_2O}。

（4）真菌和大多数细菌是还原剂，可将{CH_2O}分解为CO_2和其他简单的无机化合物。

重要名词和术语解析

Spore: All classes of microorganisms produce **spores**, metabolically inactive bodies that form and survive under adverse conditions in a "resting" state until conditions favorable for growth occur.

Reducer: Fungi, protozoa, and bacteria (with the exception of photosynthetic bacteria and protozoa) are classified as **reducers**, which break down chemical compounds to more simple species and thereby extract the energy needed for their growth and metabolism.

Producer: Algae are classified as **producers** because they utilize light energy and store it as chemical energy.

Chemotroph: **Chemotrophs** use chemical energy derived from oxidation-reduction reactions of simple inorganic chemical species for their energy needs.

Phototroph: **Phototrophs** utilize light energy from photosynthesis.

孢子：所有种类的微生物都会产生芽孢，即无代谢活性的**孢子**，在不利条件下以"静止"状态形成并存活，直至出现有利于生长的条件。

还原者：真菌、原生动物和细菌（光合作用细菌和原生动物除外）被归类为**还原者**，它们将化合物分解为更简单的物种，从而提取其生长和代谢所需的能量。

生产者：藻类被归类为**生产者**，因为它们利用光能并将其存储为化学能。

化学营养生物：**化学营养生物**使用源自简单无机化学物质的氧化还原反应的化学能满足其能量需求。

光养生物：**光养生物**利用光合作用的光能。

Heterotroph: **Heterotrophs** obtain their carbon from other organisms.

Autotroph: **Autotrophs** use carbon dioxide and ionic carbonates for the C that they require.

异养生物：异养生物从其他生物体中获取碳。

自养生物：自养生物使用二氧化碳和离子碳酸盐作为所需的碳。

3.4.2 Algae
藻类

知识框架

(1) Algae are generally microscopic organisms that subsist on inorganic nutrients using solar energy and producing biomass from carbon dioxide by photosynthesis.

1) Phycology is the study of algae.

2) Four major kinds of algae of which Chlorophyta (green algae) are responsible for most productivity in water.

(2) Main role of algae in water is production of biomass, $\{CH_2O\}$ using photochemical energy, $h\nu$:

$$CO_2 + H_2O + h\nu \longrightarrow \{CH_2O\} + O_2$$

(3) General nutrient requirements for algae:

1) Carbon from CO_2 or HCO_3^-.

2) Nitrogen, generally from NO_3^-.

3) Phosphorus from $H_2PO_4^-$ or HPO_4^{2-}.

4) Sulfur from SO_4^{2-}.

5) Trace elements (iron may be limiting). Lichen are algae and fungi in a symbiotic relationship.

Involved in rock weathering.

(1) 藻类通常是微观生物，它们利用太阳能，依靠无机养分生存，并通过光合作用从二氧化碳中产生生物质。

1) 藻类学是藻类的研究。

2) 四种主要藻类，其中 Chlorophyta（绿藻）负责水中的大部分生产力。

(2) 藻类在水中的主要作用是利用光化学能 $h\nu$ 生产生物质 $\{CH_2O\}$:

(3) 藻类的一般营养需求：

1) 来自 CO_2 或 HCO_3^- 的碳；

2) 氮，通常来自 NO_3^-；

3) 来自 $H_2PO_4^-$ 或 HPO_4^{2-} 的磷；

4) 来自 SO_4^{2-} 的硫；

5) 微量元素（铁可能是限制性的）地衣是藻类和真菌的共生关系。

参与岩石风化。

重要名词和术语解析

Algae: Algae may be considered as generally microscopic organisms that subsist on inorganic nutrients and produce organic matter from carbon dioxide by photosynthesis.

Phycology: The study of algae is called **phycology**.

Chrysophyta: **Chrysophyta** contain pigments that give these organisms a yellow-green or golden-brown color.

Diatom: **Diatoms** are characterized by silica-containing cell walls.

Chlorophyta: Commonly known as green algae, are responsible for most of the primary productivity in fresh waters.

Pyrrophyta: Commonly known as dinoflagellates, are motile with structures that enable them to move about in water.

Euglenophyta: **Euglenophyta** exhibit characteristics of both plants and animals.

Lichen: The most common symbiotic relationship involving algae is that of **lichen** in which algae coexist with fungi; both kinds of organisms are woven into the same thallus (tubular vegetative unit).

藻类:藻类通常被认为是依靠无机养分生存,并通过光合作用从二氧化碳产生有机物的微观生物。

藻类学:对藻类的研究称为藻类学。

金藻:金藻含有使生物具有黄绿色或金棕色的色素。

硅藻:硅藻的特征是细胞壁含有二氧化硅。

绿藻:通常被称为绿藻,负责淡水的大部分初级生产力。

甲藻:通常被称为鞭毛藻,具有能使它们在水中移动的结构。

裸藻:裸藻表现出植物和动物两种生物的特征。

地衣:涉及藻类的最常见的共生关系是藻类与真菌共存的**地衣**。两种生物都被编织到同一个原植体里(管状植物单元)。

3.4.3 Fungi
真菌

知识框架

Nonphotosynthetic oxic eukaryotic organisms

(1) Wide range of morphology, often filamentous.

(2) Generally do not grow well in water, but generate decomposition products that get into water.

非光合有氧真核生物

(1) 形态广泛,常呈丝状。

(2) 通常在水中生长不好,但会生成分解产物,并进入水中。

(3) Most important function in environment is breakdown of plant cellulose with *cellulase* enzyme.

(4) Key role in producing humic substance from plant matter.

(3) 环境中最重要的功能是纤维素酶分解植物纤维素。

(4) 在从植物物质生产腐殖质中的关键作用。

3.4.4 Protozoa
原生动物

知识框架

(1) Protozoa are microscopic animals composed of single eukaryotic cells.

1) Wide variety of morphology and characteristics.

2) Motility.

3) Shells.

4) Spore formation.

5) Some protozoa have chloroplasts and are photosynthetic.

(2) Protozoa are significant in water for several reasons:

1) Cause of some devastating human and animal diseases (malaria, sleeping sickness, dysentery).

2) Form mineral deposits (limestone).

3) Degrade biomass, especially in sewage treatment.

4) "Graze" on bacterial cells involved in biodegradation.

(1) 原生动物是由单个真核细胞组成的微观动物。

1) 各种各样的形态和特征。

2) 运动;

3) 壳;

4) 形成孢子;

5) 有些原生动物具有叶绿体并且具有光合作用。

(2) 原生动物在水中很重要,有以下原因:

1) 造成一些破坏性的人畜疾病(疟疾、昏睡病、痢疾);

2) 形成矿藏(石灰石);

3) 降解生物质,特别是在污水处理中;

4) 在与生物降解有关的细菌细胞上"吃草"。

3.4.5 Bacteria
细菌

知识框架

(1) Bacteria are single-celled prokaryotic

(1) 细菌是单细胞原核微

microorganisms generally in a size range of 0.5~3.0 micrometers.

1) Rods (bacillus).

2) Spheres (coccus).

3) Spirals.

4) Large surface/volume ratio makes bacteria very effective biochemical catalysts.

(2) Autotrophic and heterotrophic bacteria:

1) Autotrophic bacteria use chemical or light energy to make all required biochemical.

2) *Gallionella* oxidize iron(Ⅱ) to iron(Ⅲ) for energy:

$$4FeS(s) + 9O_2 + 10H_2O \longrightarrow 4Fe(OH)_3(s) + 4SO_4^{2-} + 8H^+$$

3) Heterotrophic bacteria metabolize organic matter.

4) Oxic (aerobic) bacteria require O_2.

5) Anoxic (anaerobic) bacteria.

6) Facultative bacteria may use O_2 or other electron receptors.

生物，大小范围通常为0.5~3.0μm。

1) 杆状（芽孢杆菌）;

2) 球状（球菌）;

3) 螺旋状;

4) 大的表面积/体积比使细菌成为非常有效的生化催化剂。

(2) 自养和异养细菌:

1) 自养细菌利用化学或光能制造所有必需的生物化学物质;

2) 披毛菌属将二价铁氧化为三价铁以获得能量:

3) 异养细菌代谢有机物;

4) 好氧（需氧）细菌需要氧气;

5) 缺氧（厌氧）细菌;

6) 兼性细菌可能使用 O_2 或其他电子受体。

重要名词和术语解析

Bacteria: Bacteria are single-celled prokaryotic microorganisms that may be shaped as rods (**bacillus**), spheres (**coccus**), or spirals (**vibrios**, **spirilla**, **spirochetes**).

Autotrophic bacteria: Autotrophic bacteria are not dependent upon organic matter for growth and thrive in a completely inorganic medium; they use carbon dioxide or other carbonate species as a carbon source.

Heterotrophic bacteria: Heterotrophic bacteria

细菌: 细菌是单细胞的原核微生物，其形状可能是杆状（芽孢杆菌），球状（球菌）或螺旋状（弧菌、螺旋藻、螺旋体）。

自养细菌: 自养细菌不依赖有机物生长，并在完全无机的介质中繁殖。他们使用二氧化碳或其他碳酸盐种类作为碳源。

异养细菌: 异养细菌依靠有机

depend upon organic compounds, both for their energy and for the carbon required to build their biomass.

Aerobic bacteria: Aerobic bacteria require oxygen as an electron receptor.

Anaerobic bacteria: Anaerobic bacteria function only in the complete absence of molecular oxygen.

Facultative bacteria: facultative bacteria utilize free oxygen when it is available and use other substances as electron receptors (oxidants) when molecular oxygen is not available.

化合物来提供能量和产生生物质所需的碳。

需氧细菌：需氧细菌需要氧气作为电子受体。

厌氧细菌：厌氧细菌仅在完全不存在分子氧的情况下作用。

兼性细菌：兼性细菌在有游离氧时会利用自由氧，在没有分子氧时会利用其他物质作为电子受体（氧化剂）。

3.4.6 The Prokaryotic Bacterial Cell 原核细菌细胞

重要名词和术语解析

Cell wall: Bacterial cells are enclosed in a **cell wall**, which holds the contents of the bacterial cell and determines the shape of the cell. The cell wall in many bacteria is frequently surrounded by a **slime layer** (capsule).

Cell membrane: The **cell membrane** or **cytoplasmic membrane** composed of protein and phospholipid occurs as a thin layer only about 7 nanometers in thickness on the inner surface of the cell wall enclosing the cellular cytoplasm.

Mesosome: Folds in the cytoplasmic membrane called **mesosomes** serve several functions.

Pili: Hairlike **pili** on the surface of a bacterial cell enable the cell to stick to surfaces. Specialized **sex pili** enable nucleic acid transfer between

细胞壁：细菌细胞被封闭在**细胞壁**中，细胞壁容纳细菌细胞的内容物并决定细胞的形状。许多细菌的细胞壁经常被**黏液层**（胶囊）包围。

细胞膜：由蛋白质和磷脂组成的**细胞膜**或**细胞质膜**，以仅约7nm厚的薄层的形式出现在细胞壁内表面，包围细胞质。

微粒体：胞质膜上的折叠被称为**微粒体**，具有多种功能。

菌毛：细菌细胞表面的毛状菌毛使细胞能够黏附在表面。专门的**性菌毛**可以在遗传物质交

bacterial cells during an exchange of genetic material.

Flagella: Somewhat similar to pili—but larger, more complex, and fewer in number—are **flagella**, moveable appendages that cause bacterial cells to move by their whipping action. Bacteria with flagella are termed **motile**.

Cytoplasm: the medium in which the cell's metabolic processes are carried out.

Nuclear body: **Nuclear body** consists of a single DNA macromolecule that controls metabolic processes and reproduction.

Inclusion: reserve food material consisting of fats, carbohydrates, and even elemental sulfur.

Ribosome: sites of protein synthesis and which contain protein and RNA.

换期间在细菌细胞之间进行核酸转移。

鞭毛：**鞭毛**是可移动的附件，与细菌的菌毛有些相似，但更大、更复杂且数量更少，可导致细菌细胞通过鞭打动作移动。带有鞭毛的细菌被称为**运动型**。

细胞质：细胞进行代谢过程的介质。

核体：**核体**由单个 DNA 大分子组成，可控制代谢过程和繁殖。

包含物：脂肪、碳水化合物甚至元素硫组成的储备食物材料。

核糖体：蛋白质合成的位点，包含蛋白质和 RNA。

3.4.7 Kinetics of Bacterial Growth
细菌生长动力学

重要名词和术语解析

Population curve: The population size of bacteria and unicellular algae as a function of time in a growth culture is a **population curve** for a bacterial culture.

Lag phase: The first region is characterized by little bacterial reproduction and is called the **lag phase**.

Log phase: Following the lag phase comes a period of very rapid bacterial growth. This is **log phase**, or exponential phase, during which the population doubles over a regular time interval called the **generation time**.

种群曲线：在生长培养中，细菌和单细胞藻类的种群大小随时间变化是细菌培养的**种群曲线**。

滞后期：第一个区域的特征是细菌繁殖少，被称为**滞后期**。

对数期：在滞后期之后是细菌快速生长的时期。这是**对数期**，即指数期，在此期间种群在称为生成时间的规定时间间隔内翻倍。

Stationary phase: The log phase terminates and the **stationary phase** begins when a limiting factor is encountered.

Death phase: After the stationary phase, the bacteria begin to die faster than they reproduce, and the population enters the **death phase**.

静止期：遇到限制因素时，对数期终止，**静止期**开始。

死亡期：在静止期之后，细菌开始死亡的速度快于其繁殖的速度，种群进入**死亡期**。

3.4.8 Bacterial Metabolism to Get Energy and Cellular Material 通过细菌代谢获取能量和细胞物质

知识框架

Produces and changes many of the materials in water

（1）Oxic respiration has O_2 as terminal electron receptor.

（2）Anoxic respiration has species such as SO_4^{2-}, NO_3^-, HCO_3^-, as terminal electron receptors.

产生并改变水中的许多物质

（1）有氧呼吸以 O_2 为末端电子受体。

（2）缺氧呼吸具有如 SO_4^{2-}、NO_3^-、HCO_3^- 的物种作为末端电子受体。

重要名词和术语解析

Bacterial metabolism: Bacterial metabolism addresses the biochemical processes by which chemical species are modified in bacterial cells.

Catabolism and anabolism: The two major divisions of bacterial metabolism are **catabolism**, energy-yielding degradative metabolism which breaks macromolecules down to their small monomeric constituents, and **anabolism**, synthetic metabolism in which small molecules are assembled into large ones.

Aerobic respiration: If the terminal electron

细菌代谢：细菌代谢涉及细菌细胞中化学物质被改变的生物化学过程。

分解代谢和合成代谢：细菌代谢的两个主要分类分别是**分解代谢**，即将大分子分解为小的单体成分的增能降解代谢，以及将小分子组装成大分子的**合成代谢**。

有氧呼吸：如果末端电子受体

acceptor is molecular O_2, the process is **aerobic respiration**.

Anaerobic respiration: If the terminal electron acceptor is another reducible species, commonly including SO_4^{2-}, NO_3^-, HCO_3^-, or iron (Ⅲ), the process is called **anaerobic respiration**.

Psychrophilic bacteria: bacteria having temperature optima below approximately 20℃.

Mesophilic bacteria: The temperature optima of **mesophilic bacteria** lie between 20℃ and 45℃.

Thermophilic bacteria: Bacteria having temperature optima above 45℃ are called **thermophilic bacteria**.

是分子 O_2，则此过程为**有氧呼吸**。

厌氧呼吸：如果末端电子受体是另一种可还原物种，通常包括 SO_4^{2-}、NO_3^-、HCO_3^-，或三价铁，则该过程称为**厌氧呼吸**。

嗜冷细菌：最佳温度低于约 20℃ 的细菌。

中温细菌：**中温细菌**的最佳温度在 20℃ 至 45℃ 之间。

嗜热细菌：最佳温度高于 45℃ 的细菌称为**嗜热细菌**。

3.4.9 Microbial Transformations of Carbon 碳的微生物转化

知识框架

(1) Photosynthesis stores photochemical energy ($h\nu$) as chemical energy in biomass ($\{CH_2O\}$).

(2) Biomass is oxidized to release a large amount of energy per electron-mol.

1) $\frac{1}{4}\{CH_2O\} + \frac{1}{4}O_2(g) \rightarrow \frac{1}{4}CO_2 + \frac{1}{4}H_2O$, $DG^0(w) = -29.9$ kcal.

2) Release of 29.9 kcal free energy for one electron-mole of reaction.

(1) 光合作用将光化学能 ($h\nu$) 存储为生物质 ($\{CH_2O\}$) 中的化学能。

(2) 每电子摩尔生物质被氧化释放出大量能量。

1) $\frac{1}{4}\{CH_2O\} + \frac{1}{4}O_2(g) \rightarrow \frac{1}{4}CO_2 + \frac{1}{4}H_2O$, $DG^0(w) = -29.9$ kcal❶。

2) 对于一个电子摩尔反应，释放 29.9kcal❶ 自由能。

---

❶ 1kcal=4.1868kJ。

(2) Methane formation (fermentation reaction).

1) $2\{CH_2O\} \rightarrow CO_2 + CH_4$, $DG^0(w) = -5.55$ kcal/electron-mole.

(2) Increasingly used to produce methane fuel from municipal waste.

(3) Microbial metabolism of hydrocarbons important in destroying wastes such as spilled crude oil.

（2）甲烷生成（发酵反应）。

1) $2\{CH_2O\} \rightarrow CO_2 + CH_4$，$DG^0(w) = -5.55$ 千卡/电子摩尔。

2）越来越多地用于从城市垃圾中生产甲烷燃料。

（3）碳氢化合物的微生物代谢对破坏废物（如溢漏的原油）非常重要。

重要名词和术语解析

Fermentation reaction: a redox process in which both the oxidizing agent and reducing agent are organic substances.

发酵反应：氧化剂和还原剂均为有机物质的氧化还原过程。

3.4.10 Biodegradation of Organic Matter 有机物的生物降解

知识框架

(1) Biodegradation of pesticides, petroleum wastes, other organic matter is very important in the environment.

(2) Enzymatic oxidation including epoxidation of aromatics.

(3) Beta-oxidation of hydrocarbon chains:

1) $CH_2CH_2CH_2CH_2CO_2H + 3CO_2 \rightarrow CH_2CH_2CO_2H + 2CO_2 + 2H_2O$.

2) Branched chains inhibitory, especially quaternary carbon.

（1）农药、石油废料和其他有机物的生物降解在环境中非常重要。

（2）酶促氧化包括芳族化合物的环氧化。

（3）烃链的 β-氧化：

1) $CH_2CH_2CH_2CH_2CO_2H + 3CO_2 \rightarrow CH_2CH_2CO_2H + 2CO_2 + 2H_2O$；

2）支链抑制，尤其是季碳。

Biodegradation of Organics Other Than Oxidation

(1) Hydrolysis, such as of insecticidal malathion:

$$(CH_3O)_2-\overset{\overset{S}{\|}}{P}-S-\overset{\overset{H}{|}}{\underset{\underset{H}{|}}{C}}-\overset{\overset{O}{\|}}{C}-O-C_2H_5 \xrightarrow{H_2O} (CH_3O)_2-\overset{\overset{S}{\|}}{P}-SH + HO-\overset{\overset{H}{|}}{\underset{\underset{H}{|}}{C}}-\overset{\overset{O}{\|}}{C}-O-C_2H_5$$

(2) Reductions such as aldehydes to alcohols.

(3) Dehalogenation, especially removal of bound Cl.

(4) Dealkylation, removal of —CH_3 from N, S, or O.

除氧化作用外有机物的生物降解

(1) 水解,例如杀虫剂马拉硫磷:

(2) 醛类还原为醇类。

(3) 脱卤,特别是去除结合的 Cl。

(4) 脱烷基,从 N、S 或 O 中除去 —CH_3 基。

重要名词和术语解析

Oxidation: **Oxidation** occurs by the action of oxygenase enzymes.

Epoxidation: **Epoxidation** consists of adding an oxygen atom between two C atoms in an unsaturated system.

Hydroxylation: **Hydroxylation** often accompanies microbial oxidation.

Hydrolysis: **Hydrolysis** involves the addition of H_2O to a molecule accompanied by cleavage of the molecule into two products, is a major step in microbial degradation of many pollutant compounds, especially pesticidal esters, amides, and organophosphate esters.

Hydrolase enzyme: The types of enzymes that bring about hydrolysis are **hydrolase enzymes**.

Esterase: Those that enable the hydrolysis of esters are called **esterases**.

氧化:氧化是通过加氧酶的作用发生的。

环氧化:环氧化是在不饱和体系的两个 C 原子之间添加一个氧原子。

羟基化:微生物氧化经常伴随着**羟基化**。

水解:水解涉及在分子中添加 H_2O,同时将分子裂解成两种产物,是许多污染物化合物(尤其是农药酯、酰胺和有机磷酸酯)微生物降解的重要步骤。

水解酶:引起水解的酶是**水解酶**。

酯酶:那些能使酯水解的酯酶称为酯酶。

Amidase: those that hydrolyze amides are **amidases**.

Reductase enzyme: **Reductions** are carried out by **reductase enzymes**; for example, nitroreductase enzyme catalyzes the reduction of the nitro group.

Dehalogenation: **Dehalogenation** reactions of organohalide compounds involve the bacterially-mediated replacement of a covalently-bound halogen atom (F, Cl, Br, I) with —OH.

Ring cleavage: **Ring cleavage** is a crucial step in the ultimate degradation of organic compounds having aryl rings.

Dealkylation: replacement of alkyl groups by H.

酰胺酶：水解酰胺的是**酰胺酶**。

还原酶：还原是通过**还原酶**进行的；例如，硝基还原酶催化硝基的还原。

脱卤：有机卤化物的**脱卤**反应涉及通过细菌介导的—OH 取代共价键合的卤素原子（F、Cl、Br、I）。

环裂解：**环裂解**是最终降解具有芳基环的有机化合物的关键步骤。

脱烷基：用 H 取代烷基。

3.4.11 Microbial Transformations of Nitrogen 氮的微生物转化

知识框架

(1) Microbial nitrogen fixation (*Rhizobium*) cyanobacteria：

$$3\{CH_2O\} + 2N_2 + 3H_2O + 4H^+ \longrightarrow 3CO_2 + 4NH_4^+$$

(2) Nitrification producing nitrate utilizable by algae, plants：

$$2O_2 + NH_4^+ \longrightarrow NO_3^- + 2H^+ + H_2O$$

(3) Nitrate reduction using NO_3^- as an electron receptor：

1) $2NO_3^- + \{CH_2O\} \rightarrow 2NO_2^- + H_2O + CO_2$.

2) Denitrification, return of N_2 to the atmosphere.

3) $4NO_3^- + 5\{CH_2O\} + 4H^+ \rightarrow 2N_2(g) + 5CO_2 + 7H_2O$.

(1) 微生物固氮蓝细菌：

(2) 硝化作用产生的硝酸盐可被藻类、植物利用：

(3) 使用 NO_3^- 作为电子受体的硝酸盐还原：

1) $2NO_3^- + \{CH_2O\} \rightarrow 2NO_2^- + H_2O + CO_2$；

2) 反硝化，以 N_2 返回大气；

3) $4NO_3^- + 5\{CH_2O\} + 4H^+ \rightarrow 2N_2(g) + 5CO_2 + 7H_2O$。

重要名词和术语解析

Nitrogen fixation: The overall microbial

固氮：固氮的整个微生物过程

process for **nitrogen fixation**, the binding of atmospheric nitrogen in a chemically combined form.

Nitrification: the conversion of N(−Ⅲ) to N(Ⅴ), is a very common and extremely important process in water and in soil.

Nitrate reduction: **Nitrate reduction** refers to microbial processes by which nitrogen in chemical compounds is reduced to lower oxidation states.

Denitrification: An important special case of nitrate reduction is **denitrification**, in which the reduced nitrogen product is a nitrogen-containing gas, usually N_2.

是以化合形式结合大气中的氮。

硝化作用：−3 价 N 转化为正 5 价 N 是水和土壤中非常普遍且极为重要的过程。

硝酸盐还原：硝酸盐还原是指将化合物中的氮还原为较低氧化态的微生物过程。

反硝化作用：硝酸盐还原的一个重要特殊情况是**反硝化作用**，其中还原后的氮产物为含氮气体，通常为 N_2。

3.4.12 Microbial Transformations of Phosphorus and Sulfur 磷和硫的微生物转化

知识框架

Phosphorus

Microbial transformations of phosphorothionate and phosphorodi-thioate ester insecticides (hydrolysis) where R is an alkyl group, Ar is a substituent group that is frequently aromatic, and X is either O or S:

磷

硫代磷酸酯和二硫代磷酸酯杀虫剂的微生物转化（水解），其中 R 为烷基，Ar 为经常为芳族的取代基，X 为 O 或 S：

$$\begin{array}{c} X \\ \parallel \\ R-O-P-OAr \\ | \\ O \\ | \\ R \end{array} \xrightarrow{H_2O} \begin{array}{c} X \\ \parallel \\ R-O-P-OH+HOAr \\ | \\ O \\ | \\ R \end{array}$$

$$\begin{array}{c} X \\ \parallel \\ R-O-P-SAr \\ | \\ O \\ | \\ R \end{array} \xrightarrow{H_2O} \begin{array}{c} X \\ \parallel \\ R-O-P-OH+HSAr \\ | \\ O \\ | \\ R \end{array}$$

Microbial reduction of sulfate (*Desulfovibrio*)　　微生物还原硫酸盐(脱硫弧菌)

$$SO_4^{2-} + 2\{CH_2O\} + 2H^+ \longrightarrow H_2S + 2CO_2 + 2H_2O$$

(1) Some bacteria (*Thiobacillus thioxidans*) oxidize sulfide to sulfuric acid:

$$H_2S \longrightarrow 2H^+ + SO_4^{2-}$$

(1) 一些细菌（硫杆菌）将硫化物氧化成硫酸：

(2) Organosulfur compounds include a variety of sulfur-containing functional groups that are converted to sulfide and eventually sulfate by biodegradation.

(2) 有机硫化合物包括各种含硫官能团，这些基团可通过生物降解转化为硫化物，最终转化为硫酸盐。

Example: Sulfur-containing amino acid cysteine.

例：含硫的氨基酸半胱氨酸。

$$\underset{\substack{H\ H\\|\ |\\H\ NH_3^+}}{HS-C-C-CO_2 + H_2O} \xrightarrow[\text{胱氨酸脱硫水化酶}]{\text{Bacteria 细菌}\atop \text{Cysteine desulfhydrase}} H_3C-\underset{O}{\overset{O}{C}}-\underset{O}{\overset{O}{C}}-OH + H_2 + NH_3$$

重要名词和术语解析

Phosphorothionate and **phosphorodithioate**: The organophosphorus compounds of greatest environmental concern tend to be sulfur-containing **phosphorothionate** and **phosphorodithioate** ester insecticides with the general formulas, where R and R′ represents a hydro-carbon substituted hydrocarbon moieties.

磷硫代磷酸酯和二硫代磷酸酯：对环境最受关注的有机磷化合物趋向于具有通式的含硫**磷硫代磷酸酯和二硫代磷酸酯**杀虫剂，其中 R 和 R′代表烃取代的烃基。

3.4.13 Microbial Transformations of Halogens and Organohalides
卤素和有机卤化物的微生物转化

知识框架

(1) Dehalogenation reactions carried out by cometabolism:

(1) 通过新陈代谢进行的脱卤反应：

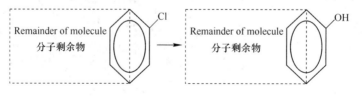

(2) Dehalorespiration reactions in which some anoxic bacteria dechlorinate chlorinated hydrocarbons by replacing Cl by H:

(2) 脱氧呼吸反应，其中一些缺氧细菌通过用 H 代替 Cl 来脱除氯代烃：

$$CH_2O+H_2O+2Cl\text{-}R \longrightarrow CO_2+2H^++2Cl^-+2H\text{-}R$$

(3) Microbial dechlorination of 1, 1, 2, 2-tetrachloroethane:

(3) 1, 1, 2, 2-四氯乙烷的微生物脱氯：

Vinyl chloride 氯乙烯　Ethene(ethylene) 乙烯

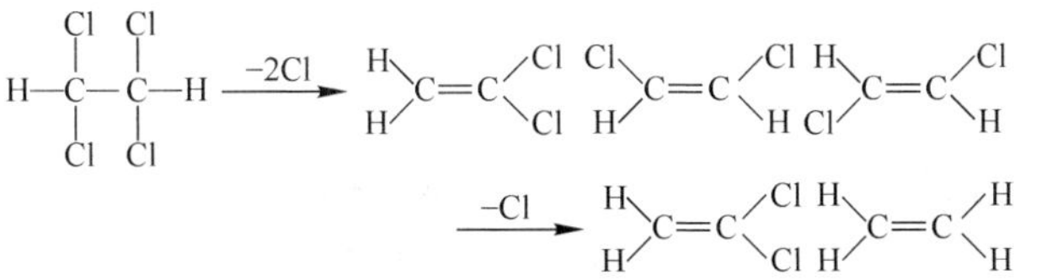

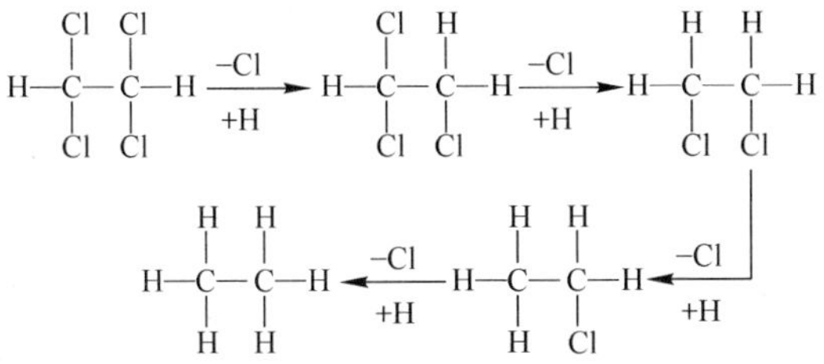

重要名词和术语解析

<u>Cometabolism</u>: Cometabolism results from a lack of specificity in the microbial degradation processes.

<u>共代谢</u>: 共代谢是由于微生物降解过程缺乏特异性所致。

3.4.14 Microbial Transformations of Metals and Metalloids 金属和准金属的微生物转化

知识框架

Bacteria that get energy by catalyzing oxidation of iron(Ⅱ) to iron(Ⅲ)

通过催化二价铁氧化为三价铁获得能量的细菌

(1) *Ferrobacillus*, *Gallionella*, some *Sphaerotilus*.

(1) 铁杆菌、披毛菌、一些球菌。

(2) Example: $4FeCO_3(s)+O_2+6H_2O \rightarrow 4Fe(OH)_3(s)+4CO_2$.

(2) 例：$4FeCO_3(s)+O_2+6H_2O \rightarrow 4Fe(OH)_3(s)+4CO_2$。

(3) Energy yield is low so production of a

(3) 产能低，因此生产少

small amount of biomass produces a large amount of solid $Fe(OH)_3$.

Acid Mine Waters

(1) Mineral pyrite, FeS_2, exposed by coal mining is oxidized largely by bacterial action to produce sulfuric acid and acidic Fe(Ⅲ) ion beginning with:

$$2FeS_2(s) + 2H_2O + 7O_2 \longrightarrow 4H^+ + 4SO_4^{2-} + 2Fe^{2+}$$
$$4Fe^{2+} + O_2 + 4H^+ \longrightarrow 4Fe^{3+} + 2H_2O$$

(2) Several kinds of bacteria may be involved:

1) *Thiobacillus ferrooxidans*.

2) *Metallogenium*.

3) *Thiobacillus thiooxidans*.

4) *Ferrobacillus ferrooxidans*.

5) Because of production of acid, bacteria involved must be very acid-tolerant.

(3) An adverse effect of acid mine water is precipitation of amorphous, semigelatinous hydrated iron(Ⅲ) oxide.

Microbial Transitions of Selenium

(1) Selenium is a nutritionally important trace element.

1) Especially for animals.

2) Selenium deficiency can be harmful.

3) Excess selenium can be toxic.

(2) Microbial action on selenium species:

1) Reduction of SeO_3^{2-} and SeO_4^{2-} to elemental selenium.

2) Oxidation of Se(0) to SeO_3^{2-} by *Thiobacillus* and *Leptothrix*.

酸性矿井水

(1) 从煤矿开采出来的硫铁矿 FeS_2 在很大程度上通过细菌作用被氧化，生成硫酸和酸性三价铁离子：

(2) 可能涉及几种细菌：

1) 氧化亚铁硫杆菌；

2) 生金菌；

3) 硫氧化硫杆菌；

4) 铁氧化铁杆菌；

5) 由于产生酸，因此涉及的细菌必须非常耐酸。

(3) 酸性矿井水的不利影响是沉淀出无定形的半胶状水合氧化铁(Ⅲ)。

硒的微生物转化

(1) 硒是重要的营养微量元素。

1) 特别是对于动物；

2) 硒缺乏可能有害；

3) 过量的硒可能有毒。

(2) 微生物对硒的作用：

1) 将 SeO_3^{2-} 和 SeO_4^{2-} 还原为元素硒；

2) 硫杆菌和纤毛菌将Se(0)氧化为 SeO_3^{2-}。

(3) Dimethyl selenide, $(CH_3)_2Se$, is a volatile selenium compound emitted to the atmosphere by bacterial action.

Probably by bacterial action on selenomethionine.

(3) 硒化二甲酯 $(CH_3)_2Se$ 是通过细菌作用释放到大气中的挥发性硒化合物。

可能是通过细菌对硒代蛋氨酸的作用。

3.5 Water Pollutants and Water Pollution 水污染物和水污染

3.5.1 Nature and Types of Water Pollutants 水污染物的性质和类型

知识框架

(1) Markers of water pollution that show presence of pollution sources.

1) Herbicides indicate agricultural runoff.

2) Fecal coliform bacteria indicate sewage sources.

3) Pharmaceutical metabolites in domestic wastewater.

(2) Biomarkers of water pollution are organisms that indicate pollution.

1) May accumulate pollutants that appear in analysis.

2) May show effects from pollutant exposure.

3) Fish lipid tissue accumulates persistent organic pollutants.

4) Osprey at top of aquatic food web is a good biomarker.

(1) 表明存在污染源的水污染标记。

1) 除草剂表明农业径流；

2) 大肠菌类细菌指示污水源；

3) 生活污水中的药物代谢物。

(2) 水污染的生物标记是表明污染的生物。

1) 可能会积累分析中出现的污染物；

2) 可能显示污染物暴露的影响；

3) 鱼的脂质组织积累了持久性有机污染物；

4) 鱼鹰在水生食物网的顶部，是很好的生物标记。

3.5.2 Elemental Pollutants
元素污染物

知识框架

Trace elements (harmful at a few parts per million or less)

(1) Heavy metals are among most harmful.
1) Cadmium.
2) Mercury.
3) Lead.
4) Most are sulfur seekers.

(2) Metalloids may be significant water pollutants.
1) Most important is arsenic.
2) Selenium and antimony can also be harmful.

痕量元素（含量为每百万分之几或更少的有害成分）

(1) 重金属是最有害的。
1) 镉；
2) 汞；
3) 铅；
4) 大多数对硫具有极大的亲和力。

(2) 类金属可能是重要的水污染物。
1) 最重要的是砷。
2) 硒和锑也可能有害。

重要名词和术语解析

Trace element: Trace element is a term that refers to those elements that occur at very low levels of a few parts per million or less in a given system.

Trace substance: The term **trace substance** is a more general one applied to both elements and chemical compounds.

Heavy metal: Some of the **heavy metals** are among the most harmful of the elemental pollutants and are of particular concern because of their toxicities to humans.

Metalloid: elements on the borderline between metals and non-metals, are significant water pollutants.

Inorganic chemical: Inorganic chemicals

痕量元素：**痕量元素**是一个术语，指的是在给定系统中以百万分之几或更少的极低水平出现的那些元素。

痕量物质：**痕量物质**是适用于元素和化合物的更通用的术语。

重金属：某些**重金属**是最有害的元素污染物，由于其对人体的毒性而特别引起关注。

类金属：金属和非金属之间的边界元素是重要的水污染物。

无机化学物质：无机化学物质

manufacture has the potential to contaminate water with trace elements.

3.5.3 Heavy Metals
重金属

知识框架

Cadmium
(1) Highly toxic.
(2) Chemically very similar to zinc.
(3) From mining and industrial wastes (especially metal plating).

Lead
(1) Widely used and distributed in the past.
(2) Plumbing (lead pipe, solder) used to be a major source.
(3) Uses (such as in gasoline) have been greatly curtailed.

Mercury
(1) Highly toxic.
(2) Mobilized by bacterial methylation—$HgCH_3^+$, $Hg(CH_3)_2$.

镉
(1) 剧毒。
(2) 化学上与锌非常相似。
(3) 来自采矿和工业废物（特别是金属电镀）。

铅
(1) 过去广泛使用和分布。
(2) 管道（铅管、焊料）曾经是主要来源。
(3) 大大减少了用途（如汽油）。

汞
(1) 剧毒。
(2) 通过细菌甲基化而增加移动性——$HgCH_3^+$、$Hg(CH_3)_2$。

重要名词和术语解析

Cadmium: Pollutant **cadmium** in water may arise from industrial discharges and mining wastes.

Lead: Inorganic **lead** arising from a number of industrial and mining sources occurs in water in the +2 oxidation state.

Mercury: Because of its toxicity, mobilization as methylated forms by anaerobic bacteria, and

镉：工业排放物和采矿废料可能产生水中的**镉**污染。

铅：多种工业和采矿来源产生的无机**铅**以正2价氧化态存在于水中。

汞：由于其毒性、由厌氧菌导致的甲基化形式的移动性，以

other pollution factors, **mercury** generates a great deal of concern as a heavy-metal pollutant.

及其他污染因素，**汞**作为重金属污染物备受关注。

3.5.4 Metalloids
类金属

知识框架

Arsenic is the most significant

(1) From coal combustion.

(2) Occurs with phosphate minerals.

(3) Byproduct of copper, gold, lead refining.

(4) Natural occurrence in some groundwaters.

(5) Formerly in pesticides: $Pb_3(AsO_4)_2$, Na_3AsO_3, $Cu_3(AsO_3)_2$.

砷是最重要的

(1) 来自煤炭燃烧。

(2) 和磷酸盐矿物质一起出现。

(3) 铜、金、铅精炼的副产品。

(4) 在某些地下水中自然发生。

(5) 以前在农药中使用: $Pb_3(AsO_4)_2$, Na_3AsO_3, $Cu_3(AsO_3)_2$。

3.5.5 Organically Bound Metals and Metalloids
有机结合的金属和准金属

知识框架

(1) Have metal (metalloid) bonded to carbon.

1) Alkyl groups such as ethyl in $Pb(C_2H_5)_4$.

2) π (pi) electron donors such as ethylene, C_2H_4.

3) Carbonyls with bound CO.

(2) Tetraethyllead, C_2H_4, in gasoline introduced large quantities of lead into the environment until it was banned.

(1) 使金属（准金属）与碳键合。

1) $Pb(C_2H_5)_4$ 中的烷基，例如乙基。

2) π(pi) 电子给体，例如乙烯 C_2H_4。

3) 结合有 CO 的羰基。

(2) 汽油中的四乙基铅 C_2H_4 将大量铅引入环境，直到被禁止。

（3）Methylation of mercury by anoxic bacteria mobilized otherwise insoluble inorganic mercury.

（4）Organotin compounds were widely used as marine biocides in ship and boat paints.

1）Tributyltin chloride commonly used industrial biocide.

2）Endocrine disruption in shellfish and oysters.

3）Uses now being phased out.

重要名词和术语解析

Carbonyls: some of which are quite volatile and toxic, having carbon monoxide bonded to metals.

羰基：其中一些具有很高的挥发性和毒性，使一氧化碳与金属结合。

3.5.6 Inorganic Species
无机物

知识框架

（1）Cyanide（HCN, CN^-）.

1）Extremely toxic.

2）Industrial uses including metal cleaning, electroplating.

3）Produced by coke ovens.

4）Water pollution and fish kills from mineral processing.

（2）Ammonia.

1）Generally as NH_4^+, NH_3 at high pH.

2）Added to drinking water for residual disinfection from chlorination.

（1）氰化物（HCN, CN^-）。

1）剧毒。

2）工业用途，包括金属清洁、电镀。

3）由焦炉生产。

4）矿物加工造成的水污染和鱼类致死。

（2）氨。

1）通常在高pH值下为NH_4^+、NH_3。

2）添加到饮用水中，以进行氯化残留消毒。

(3) Free carbon dioxide, CO_2.

1) In water from decay of organic matter, geochemical sources.

2) Makes water corrosive, harmful to aquatic life.

(4) Hydrogen sulfide, H_2S.

1) From industrial sources, decay of organosulfur compounds, geochemical sources.

2) Foul odor, very detrimental to water quality, very toxic.

3) Precipitates heavy metals.

(5) Nitrite ion, NO_2^-, intermediate in reduction of NO_3^-.

Toxic, but rare water pollutant.

(6) Sulfite ion, SO_3^{2-}.

Added to water as O_2 scavenger.

(7) Perchlorate ion, ClO_4^-.

1) Industrial pollutant in some cases.

2) Recognized as a pollutant fairly recently.

(8) Asbestos.

Causes cancer when inhaled, but unknown effects in water.

（3）游离二氧化碳，CO_2。

1）水中有机物腐烂，地球化学来源。

2）使水具有腐蚀性，对水生生物有害。

（4）硫化氢，H_2S。

1）来自工业，有机硫化合物的腐烂，地球化学来源。

2）异味，对水质非常有害，有毒。

3）沉淀重金属。

（5）亚硝酸根离子 NO_2^-，NO_3^- 还原的中间体。

有毒但稀有的水污染物。

（6）亚硫酸根离子，SO_3^{2-}。

加入水中作为氧气清除剂。

（7）高氯酸根离子 ClO_4^-。

1）在某些情况下为工业污染物。

2）最近被公认为污染物。

（8）石棉。

吸入会引起癌症，但在水中会产生未知的影响。

重要名词和术语解析

Cyanide: a deadly poisonous substance, exists in water as HCN, a weak acid, K_a of $6×10^{-10}$.

Ammonia: Ammonia is the initial product of the decay of nitrogenous organic wastes, and its presence frequently indicates the presence of such wastes.

氰化物：一种致命的有毒物质，在水中以 HCN 的形式存在，K_a 为 $6×10^{-10}$，是一种弱酸。

氨：氨是含氮有机废物腐烂的最初产物，其经常表明存在此类废物。

Hydrogen sulfide: H_2S, is a product of the anaerobic decay of organic matter containing sulfur.

Free carbon dioxide: CO_2, is frequently present in water at high levels due to decay of organic matter.

Nitrite ion: NO_2^-, occurs in water as an intermediate oxidation state of nitrogen over a relatively narrow pE range.

Sulfite ion: SO_3^{2-}, is found in some industrial wastewaters.

硫化氢：H_2S，是含硫有机物的厌氧分解产物。

游离二氧化碳：CO_2，由于有机物的腐烂，水中的二氧化碳含量很高。

亚硝酸根离子：NO_2^-，在水中以相对较窄的 pE 范围内的氮的中间氧化态存在。

亚硫酸根离子：SO_3^{2-}，在一些工业废水中发现亚硫酸根离子。

3.5.7 Algal Nutrients and Eutrophication 藻类营养物和富营养化

知识框架

（1）Eutrophication means "well nourished".

（2）Eutrophication in excess is detrimental causing heavy growth of biomass followed by decay.
1）Consumes O_2.
2）Fills shallow water bodies.

（3）Of numerous algal nutrients phosphorus is generally limiting and is limited to control eutrophication.

（1）富营养化意味着"营养丰富"。

（2）过度富营养化有害，导致生物量大量增长，继而腐烂。
1）消耗氧气；
2）填充浅水区。

（3）在众多藻类营养物中，磷通常是限制性的，并且受限于控制富营养化。

重要名词和术语解析

Eutrophication: The term **eutrophication**, derived from the Greek word meaning "well-nourished", describes a condition of lakes or reservoirs involving excess algal growth.

富营养化：富营养化一词源自希腊语，意思是"营养丰富"，描述了湖泊或水库中藻类过度生长的情况。

3.5.8 Acidity, Alkalinity and Salinity
酸度、碱度和盐度

知识框架

(1) Acid.
1) Pollutant acids generally strong acids.
2) One of the most common is acid mine water (H_2SO_4).
3) Potential industrial sources of pollution.
(2) Alkalinity.
1) Generally due to $NaHCO_3$.
2) From natural geological sources.
3) Can be worsened by irrigation practices.
(3) Salinity.
1) Salts such as NaCl and Na_2SO_4.
2) Increased in municipal water systems.
3) Increased by irrigation.
4) Major problem, especially in heavily irrigated areas.

(1) 酸。
1) 污染物酸通常是强酸；
2) 最常见的一种是酸性矿水(H_2SO_4)。
3) 潜在的工业污染源。
(2) 碱度。
1) 通常是由于$NaHCO_3$；
2) 来自自然地质资源；
3) 灌溉习惯会使情况恶化。
(3) 盐度。
1) 盐，例如 NaCl 和 Na_2SO_4；
2) 市政供水系统增加；
3) 通过灌溉增加；
4) 重大问题，特别是在灌溉严重的地区。

重要名词和术语解析

Pollutant acid: The most common source of **pollutant acid** in water is acid mine drainage.
Alkalinity: Excess **alkalinity**, and frequently accompanying high pH, generally are not introduced directly into water from anthropogenic sources.

污染物酸：水中最常见的**污染物酸**来源是酸性矿山排水。
碱度：过量的**碱度**，通常伴随着高 pH 值，不会从人为来源直接引入水中。

3.5.9 Oxygen, Oxidants and Reductants
氧、氧化剂和还原剂

知识框架

Dissolved oxygen, DO, is important in water
(1) Depleted by oxidation of NH_4^+, Fe^{2+},

溶解氧 DO 在水中很重要
(1) 通过 NH_4^+、Fe^{2+}、SO_3^{2-}

SO_3^{2-}, and especially biodegradation of biomass, {CH_2O}:

$$\{CH_2O\} + O_2 \longrightarrow CO_2 + H_2O$$

(2) Biochemical oxygen demand, BOD, refers to amount of oxygen consumed in a volume of water by the biodegradable organic matter in it.

(3) Total organic carbon, TOC, is often substituted for BOD but measures non-biodegradable organic matter as well.

重要名词和术语解析

Biochemical oxygen demand: The degree of oxygen consumption by micro-bially-mediated oxidation of con-taminants in water is called the **biochemical oxygen demand** (or biological oxygen demand), **BOD**.

的氧化而耗竭，尤其是生物质 {CH_2O} 的生物降解。

(2) 生化需氧量 BOD，是指一定量的水被其中可生物降解的有机物消耗的氧气量。

(3) 总有机碳 TOC 通常可以代替生化需氧量，但也可以测量不可生物降解的有机物。

生化需氧量：通过微生物介导的污染物氧化而消耗的氧气量称为**生化需氧量**（或生物需氧量），**BOD**。

3.5.10 Organic Pollutants
有机污染物

知识框架

(1) Bioaccumulation of Organic Pollutants.

1) Bioconcentration factor (BCF) =
$$\frac{\text{Substance concentration in organism}}{\text{Substance concentration in water}}$$

2) Bioaccumulation factor, BAF, considers pollutant concentration in food as well as water.

3) Sewage.

① Contains many pollutants including pathogenic microorganisms, detergents, salts, solids.

② Most significant pollutant in sewage is biodegradable organic material ({CH_2O})

(1) 有机污染物的生物蓄积。

1) 生物浓缩系数（BCF）=
$$\frac{\text{生物体中的物质浓度}}{\text{水中物质浓度}}$$

2) 生物富集系数 BAF 考虑食物和水中的污染物浓度。

3) 污水。

①包含许多污染物，包括病原微生物、洗涤剂、盐、固体；

②污水中最重要的污染物是可生物降解的有机物质

manifested as biochemical oxygen demand (BOD).

③Main objective of wastewater treatment is elimination of BOD.

Soaps, Detergents, and Detergent Builders

(1) Soaps are salts of long-chain fatty acids.

1) Sodium stearate: $C_{17}H_{35}CO_2^-Na^+$.

2) Soaps form spherical micelles which may entrain water-insoluble grease and oil particles.

3) Soap lowers water surface tension which aids its cleaning action.

4) Soaps are biodegradable.

5) Soaps produce insoluble salts with divalent metal ions, predominantly calcium, which removes them from water, but reduces their effectiveness as cleaning agents in hard water.

(2) Calcium stearate:
$$Ca^{2+}(C_{17}H_{35}CO_2^-)_2(s)$$

Detergents

(1) Synthetic detergents lower water surface tension and enable its cleaning action.

1) Do not form precipitates with hardness ions.

2) Amphiphilic structure with ionic "head" and hydrocarbon "tail".

3) Detergent surfactants concentrate at interfaces of water with air, solids (dirt), and immiscible greases and oils.

(2) Poorly biodegradable ABS surfactants formerly used.

({CH_2O}),表现为生化需氧量(BOD);

③废水处理的主要目标是消除生化需氧量。

肥皂、洗涤剂和助洗剂

(1) 肥皂是长链脂肪酸的盐。

1) 硬脂酸钠: $C_{17}H_{35}CO_2^-Na^+$。

2) 肥皂形成球形胶束,可能夹带不溶于水的油脂和油粒。

3) 肥皂降低了水的表面张力,有助于清洁。

4) 肥皂是可生物降解的。

5) 肥皂会与二价金属离子(主要是钙)产生不溶性盐,将其从水中清除,但会降低其在硬水中作为清洁剂的功效。

(2) 硬脂酸钙:

洗涤剂

(1) 合成洗涤剂可降低水表面张力并使其具有清洁作用。

1) 不与硬离子形成沉淀;

2) 具有离子"头"和碳氢化合物"尾"的两亲结构;

3) 洗涤剂表面活性剂集中在水与空气、固体(污垢)以及不混溶的油脂和油的界面上。

(2) 以前使用的生物降解性差的 ABS 表面活性剂。

(3) Biodegradable LAS surfactants now used.

Alkyl polyethoxylate surfactants（structural formula below）

$$HO-\underset{H}{\overset{H}{C}}-\underset{H}{\overset{H}{C}}-O\left[\right]_n-C_6H_4-C-C-C-C-C-C-C-C-H$$

Nonylphenol polyethoxylate
甲基苯酚聚乙二醇

(1) Used as detergents, dispersing agents, emulsifiers, solubilizers, wetting agents.
(2) Resist biodegradation.
(3) Xenoestrogens of health concern.

Detergent formulations have numerous components.
(1) Examples: Alkalies, anticorrosive silicates.
(2) Builders added to improve performance have caused problems.
(3) Polyphosphates used in builders hydrolyze to phosphates that cause eutrophication.

Naturally Occurring Chlorinated and Brominated Compounds
(1) Produced mostly by marine organisms. Chemical defense agents.
(2) Detected in Arctic samples:
1) Air.
2) Fish.
3) Seabird eggs.
4) Marine mammals.
5) Human milk.

(3) 现在使用可生物降解的 LAS 表面活性剂。

烷基聚乙氧基化物表面活性剂（以下结构式）

(1) 用作洗涤剂、分散剂、乳化剂、增溶剂、湿润剂。
(2) 抵抗生物降解。
(3) 健康方面的异雌激素。

洗涤剂配方具有许多成分。
(1) 例：碱、防腐硅酸盐。
(2) 为提高性能而添加的助洗剂引起了问题。
(3) 助洗剂中使用的多磷酸盐水解成引起富营养化的磷酸盐。

自然发生的氯化和溴化化合物
(1) 主要由海洋生物生产。化学防御剂。
(2) 在北极样品中检测到：
1) 空气；
2) 鱼；
3) 海鸟卵；
4) 海洋哺乳动物；
5) 母乳。

(3) Example below:

1,2′-Bi-1H-pyrrole,2,3,3′,4,.4′,5,5′,-heptachloro-1′methyl
1,2′-双-1H-吡咯,2,3,3′,4,.4′,5,5′,-七氯-1′甲基

(3) 下面的例子:

Microbial Toxins in Water

(1) Toxins from aquatic bacteria and protozoa.

(2) Many, such as *Cylindrospermopsin* from cyanobacteria.

(3) Protozoal dinoflagellata toxins cause many maladies.

1) Gastrointestinal, respiratory, skin disorders in humans.

2) Mass kills of marine mammals.

3) Sometimes fatal paralytic conditions from eating infested shellfish.

(4) Red tides from explosive growth of dinoflagellates.

1) Turn water red, yellow, olive-green.

2) Kill marine organisms.

3) Make sea spray very irritating to humans.

水中的微生物毒素

(1) 来自水生细菌和原生动物的毒素。

(2) 有许多,比如来自蓝细菌的柱孢藻毒素。

(3) 原生动物鞭毛虫毒素引起许多疾病。

1) 人类的胃肠道、呼吸道、皮肤疾病。

2) 大规模杀害海洋哺乳动物。

3) 有时食用被侵染的贝类会导致致命的瘫痪状况。

(4) 鞭毛虫爆发性增长引起的赤潮。

1) 将水变成红色、黄色、橄榄绿色。

2) 杀死海洋生物。

3) 使海浪对人体非常刺激。

重要名词和术语解析

Soap: **Soaps** are salts of higher fatty acids, such as sodium stearate, $C_{17}H_{35}COO^-Na^+$.

Detergent: Synthetic **detergents** have good

肥皂: **肥皂**是高级脂肪酸的盐,比如硬脂酸钠 $C_{17}H_{35}COO^-Na^+$。

洗涤剂: 合成**洗涤剂**具有良好

cleaning properties and do not form insoluble salts with "hardness ions" such as calcium and magnesium.

Surfactant: The key ingredient of detergents is the **surfactant** or surface-active agent, which acts in effect to make water "wetter" and a better cleaning agent.

Amphiphilic structure: one part of the molecule is a polar or ionic group (head) with a strong affinity for water, and the other part is a hydrocarbon group (tail) with an aversion to water.

Biorefractory organic: **Biorefractory organics** are the organic compounds of most concern in wastewater, particularly when they are found in sources of drinking water.

的清洁性能，并且不会与"硬度离子"（如钙和镁）形成不溶性盐。

表面活性剂：清洁剂的关键成分是**表面活性剂**，其作用是使水"更湿"并成为更好的清洁剂。

两亲结构：分子的一部分是对水具有强亲和力的极性或离子基团（头），另一部分是具有憎水性的烃基团（尾）。

生物难降解有机物：**生物难降解有机物**是废水中最受关注的有机化合物，尤其是在饮用水源中发现它们时。

3.5.11 Pesticides in Water
水中的农药

知识框架

(1) Numerous kinds of compounds added to soil and plants.

1) Insecticide.
2) Herbicides.
3) Molluscicides.
4) Fungicides.
5) Bactericides.
6) Slimicides.
7) Avicides (birds).
8) Piscicides (fish).
9) Plant growth regulators.
10) Plant defoliants.
11) Plant desiccants.

(1) 添加到土壤和植物中的多种化合物。

1) 杀虫剂；
2) 除草剂；
3) 杀软体动物剂；
4) 杀真菌剂；
5) 杀细菌剂；
6) 杀黏菌剂；
7) 驱鸟剂（鸟）；
8) 灭鱼药（鱼）；
9) 植物生长调节剂；
10) 植物脱叶剂；
11) 植物干燥剂。

(2) Insecticides and fungicides most important for human exposure because of application near harvest time.

(3) Herbicides most important water pollutants because of widespread application directly onto soil.

Pesticides and Other Chemicals of Particular Concern for Water Pollution

(1) Highly biodegradation resistant compounds.

(2) Known or probable carcinogens.

(3) Substances with adverse reproductive or developmental effects.

(4) Neurotoxins including cholinesterase inhibitors.

(5) Acutely toxic substances.

(6) Known groundwater contaminants.

Natural Product Insecticides

(1) Nicotine from tobacco.

(2) Rotenone from some legume roots.

Pyrethrins

Pyrethroids are synthetic analogs of pyrethrins.

(1) Allethrin.

(2) Fenvalerate.

(3) Cypermethrin.

The Emergence of Neonicotinoid Insecticides

(1) Neonicotinoid insecticides since shortly before.

（2）杀虫剂和杀真菌剂对人类暴露最重要，因为在收获期附近施用。

（3）除草剂是最重要的水污染物，因为直接广泛应用于土壤。

特别关注水污染的农药和其他化学品

（1）高度抗生物降解的化合物。

（2）已知或可能的致癌物。

（3）对生殖或发育有不利影响的物质。

（4）神经毒素，包括胆碱酯酶抑制剂。

（5）剧毒物质。

（6）已知的地下水污染物。

天然杀虫剂

（1）烟草中的尼古丁。

（2）一些豆科植物根中的鱼藤酮。

除虫菊酯

拟除虫菊酯是除虫菊酯的合成类似物。

（1）丙烯菊酯。

（2）氰戊菊酯。

（3）氯氰菊酯。

新烟碱类杀虫剂的出现

（1）1990 自 1990 年之前不久以来的新烟碱类杀虫剂。

(2) Loosely related to natural product botanical nicotine.

(3) Emerged as the most widely used insecticides worldwide.

1) Annual production of several thousand metric tons.

2) About one third of world insecticide market.

3) Uses include protection of crops, fruits and vegetables, pest control in fish farming, veterinary products, and other uses.

4) Concerns regarding honeybee colony collapse.

5) Restricted in Europe in 2013 on crops judged to be attractive to bees.

Organochlorine insecticides have been mostly phased out of use because of persistence, harm to birds.

Endosulfan one of the last to be eliminated.

Organophosphate insecticides

Though biodegradable and not generally environmentally harmful, some very toxic effects (acetycholinesterase inhibitors).

Malathion is a commonly used organophosphate because mammals can hydrolyze it to non-toxic products.

Carbamates: Esters of Carbamic Acid

(1) Biodegradable.

(2) Acetylcholinesterase inhibitors but not unduly toxic.

（2）与天然产物植物烟碱松散相关。

（3）成为全球使用最广泛的杀虫剂。

1）年产量数千吨。

2）约三分之一的世界杀虫剂市场。

3）用途包括保护农作物、水果和蔬菜、鱼类养殖中的病虫害防治、兽医产品以及其他用途。

4）有关蜜蜂群落崩溃的担忧。

5）2013年在欧洲被限制使用在对蜜蜂具有吸引力的农作物。

由于有机氯杀虫剂的持久性、对鸟类的伤害，大部分已被淘汰。

硫丹是最后被淘汰的。

有机磷杀虫剂

尽管可生物降解，并且通常对环境无害，但有一些剧毒作用（乙酰胆碱酯酶抑制剂）。

马拉硫磷是一种常用的有机磷酸盐，因为哺乳动物可以将其水解为无毒产品。

氨基甲酸酯

（1）可生物降解。

（2）乙酰胆碱酯酶抑制剂，但毒性不大。

Fungicides

Herbicides

(1) Herbicides applied to millions of acres of cropland worldwide to control weeds.

(2) Herbicides commonly occur in surface water and groundwater, especially in intensely cropped areas.

1) Especially common are atrazine, simazine, and cyanazine used for weed control on corn and soybeans.

2) Although widely used to control weeds on crops genetically engineered to resist it, glyphosate has a very strong affinity for soil and rarely gets into water.

Paraquat, used since 1965, has caused many deaths because of its high toxicity.

2,4,5-T ("Agent Orange") no longer used.

Arsenic compounds were widely used for pesticides in the past and still contaminate soil in some areas such as sites of old orchards.

杀真菌剂

除草剂

(1) 除草剂已应用于全球数百万英亩的农田以控制杂草。

(2) 除草剂通常存在于地表水和地下水中，特别是在种植面积大的地区。

1) 特别常见的是用于抑制玉米和大豆杂草的阿特拉津、西马津和氰嗪。

2) 尽管草甘膦广泛用于控制转基因作物的杂草，但它对土壤具有很强的亲和力，很少溶于水。

自1965年以来使用的百草枯，由于其高毒性而导致许多人死亡。

不再使用2,4,5-T（"橙色代理商"）。

过去，砷化合物被广泛用作农药，但在某些地区，如老的果园，仍然污染土壤。

重要名词和术语解析

Insecticides and **molluscicides**: For the control of snails and slugs.
Nematicides: For the control of microscopic roundworms.
Rodenticides: Which kill rodents.
Avicides: Used to repel birds.
Piscicides: Used in fish control.
Herbicides: Used to kill plants.

杀虫剂和**杀软体动物剂**：用于控制蜗牛和鼻涕虫。
杀线虫剂：用于控制微观蛔虫。

杀鼠剂：杀死啮齿动物。
驱鸟剂：用于驱赶鸟类。
灭鱼药：用于鱼类控制。
除草剂：用于杀死植物。

Plant growth regulators, defoliants and **plant desiccants**: For various purposes in the cultivation of plants.
Fungicides: Used against fungi.
Bactericides: Against bacteria.
Slimicides: Against slime-causing organisms in water.
Algicides: Against algae.
Nicotine: From tobacco.
Rotenone: Extracted from certain legume roots, and **pyrethrins.**
Pyrethroids: Synthetic analogs of the pyrethrins, **pyrethroids**, have been widely produced as insecticides during recent years.
DDT: Dichlorodi-phenyltrichloroethane or 1,1,1-trichloro-2,2-bis(4-chlorophenyl)ethane, which was used in massive quantities following World War II.
Methoxychlor: A popular DDT substitute, reasonably biodegradable, and with a low toxicity to mammals.
Chlordane, aldrin, dieldrin/endrin and **heptachlor**: Structurally similar, all now banned for application in the U.S., share common characteristics of high persistence and suspicions of potential carcinogenicity.
Toxaphene: Toxaphene is a mixture of up to 177 individual compounds produced by chlorination of camphene, a terpene isolated from pine trees, to give a material that contains about 68% Cl and has an empirical formula of $C_{10}H_{10}Cl_8$.
Lindane: A formulation of the essentially pure

植物生长调节剂、脱叶剂和植物干燥剂：用于植物栽培的各种目的。
杀真菌剂：用于杀真菌。
杀细菌剂：针对细菌。
杀黏菌剂：针对水中的引起黏液的生物。
除藻剂：针对藻类。
尼古丁：来自烟草。
鱼藤酮：从某些豆科植物根和除虫菊酯中提取。
拟除虫菊酯：近年来，除虫菊酯的合成类似物，即拟除虫菊酯，被广泛用作杀虫剂。
DDT：二氯二苯基三氯乙烷或1,1,1-三氯-2,2-双（4-氯苯基）乙烷，在第二次世界大战后被大量使用。
甲氧基氯：一种流行的DDT替代品，可生物降解，对哺乳动物毒性低。
氯丹、艾氏剂、狄氏剂/异狄氏剂和七氯：结构相似，目前都禁止在美国使用，它们具有高持久性和疑似潜在致癌性的共同特征。
毒杀芬：毒杀芬是一种由最多177种化合物组成的混合物，这些化合物是由樟脑（一种从松树中分离出的萜烯）氯化生成的，其氯含量约为68%，经验式为$C_{10}H_{10}Cl_8$。
林丹：基本纯净的伽马异构体

gamma isomer has been marketed as the insecticide called **lindane**.

Organophosphate insecticides: insecticidal organic compounds that contain phosphorus, some of which are organic esters of orthophosphoric acid, such as paraoxon.

Carbamate: Pesticidal organic derivatives of carbamic acid.

Carbaryl: Carbaryl has been widely used as an insecticide on lawns or gardens. It has a low toxicity to mammals.

Carbofuran: Carbofuran has a high water solubility and acts as a plant systemic insecticide.

Pirimicarb: Pirimicarb has been widely used in agriculture as a systemic aphicide.

Triazine: A number of important herbicides contain three heterocyclic nitrogen atoms in ring structures and are therefore called **triazines**.

3.5.12 Organochlorine Compounds in Water
水中的有机氯化合物

知识框架

Pollutants from Pesticide Manufacture

(1) 2,3,7,8-Tetrachlorodibenzo-p-dioxin (TCDD or "dioxin"):

1) Badly contaminated some areas in 1970s.

2) Highly toxic to some animals.

(2) Kepone, manufacture of which badly contaminated the James River of Virginia in the 1970s.

的一种制剂的杀虫剂,已被称为**林丹**上市销售。

有机磷杀虫剂:含有磷的杀虫有机化合物,其中一些是正磷酸的有机酯,如对氧磷。

氨基甲酸酯:氨基甲酸的杀虫有机衍生物。

西维因:西维因在草坪或花园中被广泛用作杀虫剂。它对哺乳动物具有低毒性。

呋喃丹:呋喃丹具有很高的水溶性,可作为植物系统性杀虫剂。

灭虫威:灭虫威已被广泛用作农业上的系统性杀虫剂。

三嗪:许多重要的除草剂在结构中包含三个杂环氮原子,因此被称为**三嗪**。

农药生产中的污染物

(1) 2,3,7,8-四氯二苯并-对二噁英(TCDD或"二噁英"):

1) 在1970年代严重污染了某些地区。

2) 对某些动物有剧毒。

(2) 聚氯酮,其制造商在1970年代严重污染了弗吉尼亚州的詹姆斯河。

Polychlorinated Biphenyls (PCBs)

(1) Very high chemical, thermal, and biological stability.

(2) Had many uses, especially in electrical equipment.

(3) Highly persistent and found in water, sediments, bird and fish tissues.

(4) Contamination of Hudson River sediments from electrical equipment manufacture.

(5) Now banned:

$$Cl-\underset{\underset{H}{|}}{\overset{\overset{H}{|}}{C}}-\underset{\underset{Cl}{|}}{\overset{\overset{H}{|}}{C}}-\underset{\underset{H}{|}}{\overset{\overset{H}{|}}{C}}-Cl$$

1, 2, 3-Trichloropropane
1, 2, 3-三氯丙烷

Used as a soil fumigant.

Problem with groundwater contamination in California.

Naturally occurring chlorinated and brominated compounds produced by marine organisms.

多氯联苯（PCB）

（1）极高的化学、热和生物稳定性。

（2）有很多用途，尤其是在电气设备中。

（3）高度持久，存在于水、沉积物、鸟类和鱼类组织中。

（4）电气设备制造对哈德逊河沉积物的污染。

（5）现在被禁止：

用作土壤熏蒸剂。

加利福尼亚的地下水污染问题。

海洋生物产生的天然存在的氯化和溴化化合物。

重要名词和术语解析

Polychlorinated dibenzodioxin: The most notorious byproducts of pesticide manufacture are **polychlorinated dibenzodioxins**.

Dioxin: Of the dioxins, the most notable pollutant and hazardous waste compound is **2,3,7,8-tetrachlorodibenzo-*p*-dioxin** (**TCDD**), often referred to simply as "**dioxin**".

Polychlorinated biphenyl: polychlorinated biphenyls (**PCB** compounds) have been found throughout the world in water, sediments, bird tissue, and fish tissue.

多氯二苯并二噁英：农药生产中最臭名昭著的副产品是**多氯二苯并二噁英**。

二噁英：在二噁英中，最显著的污染物和有害废物化合物是**2,3,7,8-四氯二苯并-对二噁英**（**TCDD**），通常简称为"二噁英"。

多氯联苯：在世界各地的水、沉积物、鸟类组织和鱼类组织中都发现了**多氯联苯**（**PCB**化合物）。

3.5.13 Emerging Water Pollutants, Pharmaceuticals and Household Wastes
新兴水污染物、药品和生活垃圾

知识框架

Emerging water pollutants are relatively new substance coming into use that may get into water and that may have undiscovered pollution effects.

(1) Nanomaterials composed of very small particles with many developing uses including pharmaceuticals.

(2) Siloxanes (silicones) used in cosmetic products, water-repellant windshield coatings, detergents.

(3) Disinfection byproducts (trihalomethanes).

(4) Household wastes including surfactants, flame retardants, and plasticizers (bisphenol-A).

新兴的水污染物是相对较新的正在使用的物质，它可能会进入水中并且可能具有未发现的污染影响。

(1) 由非常小的颗粒组成的纳米材料，具有许多发展中的用途，包括药物。

(2) 用于化妆品，防水挡风玻璃涂料，洗涤剂的硅氧烷。

(3) 消毒副产物（三卤甲烷）。

(4) 家庭废物，包括表面活性剂、阻燃剂和增塑剂（双酚A）。

Pharmaceuticals and Partial Degradation Products
药品和部分降解产品

Bactericides in water
Used in soaps, shampoo, deoderants, lotions, other.

水中杀菌剂
用于肥皂、洗发水、去污剂、乳液，其他。

Estrogenic Substances in Wastewater
(1) Disrupt endocrine gland activities regulating metabolism and reproductive functions of organisms.

(2) Aquatic organisms (fish, frogs, alligators) exhibit:
1) Reproductive dysfunction.
2) Altered sex characteristic.

废水中的雌激素物质
(1) 破坏调节生物体的代谢和生殖功能的内分泌腺活动。

(2) 水生生物（鱼类、青蛙、短吻鳄）展示出：
1) 生殖功能障碍；
2) 性别特征改变；

3) Abnormal serum steroid levels.

(3) Substances include:

1) Oral contraceptives.

2) Hormonal treatments.

3) Cancer treatments.

(4) Nonionic surfactant polyethoxylates are estrogenic substances.

1) Much less potent than hormonal substances.

2) But released in enormous quantities.

Biorefractory Organic Pollutants

(1) Poorly biodegradable organics.

(2) Also called persistent organic pollutants, POP.

(3) Include prominently chlorinated hydrocarbons.

(4) Examples are benzene, chloroform, tetrachloroethylene.

Biorefractory compounds are not well removed by biological treatment and may require physical means such as carbon adsorption.

3.5.14 Radionuclides in the Aquatic Environment 水生环境中的放射性核素

知识框架

(1) Radionuclides produce ionizing radiation as alpha particles, beta particles, gamma rays.

(2) Radionuclides that may be in water.

3) 血清类固醇水平异常。

(3) 物质包括:

1) 口服避孕药;

2) 激素治疗;

3) 癌症治疗。

(4) 非离子表面活性剂聚乙氧基化物是雌激素物质。

1) 比荷尔蒙物质效力低得多。

2) 但大量释放。

生物难降解有机污染物

(1) 生物降解性差的有机物。

(2) 也称为持久性有机污染物,POP。

(3) 包括显著氯化的碳氢化合物。

(4) 例如苯、氯仿、四氯乙烯。

通过生物处理不能很好地去除生物难降解化合物,可能需要物理手段,例如碳吸附。

(1) 放射性核素以 α 粒子、β 粒子、γ 射线的形式产生电离辐射。

(2) 水中可能存在的放射性核素。

(3) Carbon-14 from cosmic processes.

(4) Radium-226 from uranium.

Especially significant water pollutant in some areas.

(5) Fission products important in health.

1) Strontium-90 that substitutes for calcium in bone.

2) Iodine-131 that affects thyroid.

3) Cesium-137 that substitutes for sodium.

(6) Radionuclides are characterized by half-lives in which half of the radioisotopes decay.

(7) Radiation from radionuclides damages living organisms by breaking bonds in biological macromolecules.

1) Bone marrow may be damaged in acute cases.

2) Genetic effects (damage to DNA) is of particular concern.

(3) 宇宙过程中的碳14。

(4) 铀中的镭226。

在某些地区特别重要的水污染物。

(5) 裂变产品对健康很重要。

1) 锶90替代骨骼中的钙；

2) 碘131影响甲状腺；

3) 铯137替代钠。

(6) 放射性核素的特征是半衰期，期间一半的放射性同位素会衰变。

(7) 放射性核素的辐射通过破坏生物大分子中的键来破坏生物体。

1) 急性情况下可能会破坏骨髓。

2) 遗传效应（对DNA的损害）特别令人关注。

重要名词和术语解析

Radionuclide：The massive production of **radionuclides**（radioactive isotopes）by weapons and nuclear reactors since World War II has been accompanied by increasing concern about the effects of radioactivity upon health and the environment.

Ionizing radiation：Radionuclides differ from other nuclei in that they emit **ionizing radiation**—alpha particles, beta particles, and gamma rays.

放射性核素：自第二次世界大战以来，武器和核反应堆大量生产**放射性核素**（放射性同位素）的同时，人们日益关注放射性对健康和环境的影响。

电离辐射：放射性核素与其他核的不同之处在于它们会发出**电离辐射**——α粒子、β粒子、γ射线。

Alpha particle: The most massive of these emissions is the **alpha particle**, a helium nucleus of atomic mass 4, consisting of two neutrons and two protons.

Beta particle: The chlorine-38 nucleus is radioactive and loses a negative **beta particle** to become an argon-38 nucleus.

Gamma ray: **Gamma rays** are electromagnetic radiation similar to X-rays, though more energetic.

Ionizing radiation: The primary effect of alpha particles, beta particles, and gamma rays upon materials is the production of ions; therefore, they are called **ionizing radiation**.

Decay: The **decay** of a specific radionuclide follows first-order kinetics.

Activity: Since the exact number of disintegrations per second is difficult to determine in the laboratory, radioactive decay is often described in terms of the **activity**, A, which is proportional to the absolute rate of decay.

Radium: The radionuclide of most concern in drinking water is **radium**, Ra.

α粒子：这些粒子中最大的是α粒子，这是原子质量为4的氦核，由两个中子和两个质子组成。

β粒子：氯38核具有放射性，失去负的**β粒子**，成为氩38核。

伽马射线：**伽马射线**是类似于X射线的电磁辐射，但能量更高。

电离辐射：α粒子、β粒子、γ射线对材料的主要作用是产生离子；因此，它们被称为**电离辐射**。

衰变：特定放射性核素的**衰变**遵循一阶动力学。

活度：由于很难在实验室中确定每秒的崩解数，因此经常用**活度** A 来描述放射性衰变，活度 A 与绝对衰变率成正比。

镭：饮用水中最受关注的放射性核素是镭 Ra。

Exercises
习题

3-1 Alkalinity is determined by titration with standard acid. The alkalinity is often expressed as mg/L of $CaCO_3$. If V_P (mL) of acid of normality N are required to titrate V_S (mL) of sample to the phenolphthalein endpoint, what is the

3-1 碱度用标准酸滴定来测定。碱度常以 mg/L 的 $CaCO_3$ 表示。如果需要 V_P（mL）的 N 正酸来将 V_S（mL）的样品滴定至酚酞终点，则酚酞碱度的计算公式是什

formula for the phenolphthalein alkalinity as mg/L of $CaCO_3$?

3-2 Exactly 100 pounds❶ of cane sugar (dextrose), $C_{12}H_{22}O_{11}$, were accidentally discharged into a small stream saturated with oxygen from the air at 25℃. How many liters of this water could be contaminated to the extent of removing all the dissolved oxygen by biodegradation?

3-3 Water with an alkalinity of $2.00×10^{-3}$ mol/L has a pH of 7.00. Calculate $[CO_2]$, $[HCO_3^-]$, $[CO_3^{2-}]$, and $[OH^-]$.

3-4 Through the photosynthetic activity of algae, the pH of the water in Problem 3-3 was changed to 10.00. Calculate all the preceding concentrations and the weight of biomass, $\{CH_2O\}$, produced. Assume no input of atmospheric CO_2.

3-5 Calcium chloride is quite soluble, whereas the solubility product of calcium fluoride, CaF_2, is only $3.9×10^{-11}$. A waste stream of $1.00×10^{-3}$ mol/L HCl is injected into a formation of limestone, $CaCO_3$, where it comes into equilibrium. Give the chemical reaction that occurs and calculate the hardness and alkalinity of the water at equilibrium. Do the same for a waste stream of $1.00×10^{-3}$ mol/L HF.

3-6 For a solution having $1.00×10^{-3}$ mol/L total alkalinity (contributions from HCO_3^-, CO_3^{2-}, and OH^- at $[H^+]=4.69×10^{-11}$), what is the percentage contribution to

3-2 100磅❶蔗糖（葡萄糖）$C_{12}H_{22}O_{11}$，在25℃时意外地被排放到氧气饱和的小溪中。现因为所有的溶解氧被生物降解过程所消耗，导致多少升水被污染？

3-3 碱度为$2.00×10^{-3}$ mol/L的水的pH值为7.00。计算$[CO_2]$、$[HCO_3^-]$、$[CO_3^{2-}]$和$[OH^-]$。

3-4 通过藻类的光合作用，将题3-3中水的pH值改为10.00，计算前面所有的浓度和产生生物质$\{CH_2O\}$的重量。假设无大气CO_2的输入。

3-5 氯化钙的溶解度相当高，而氟化钙CaF_2的溶度积只有$3.9×10^{-11}$。将$1.00×10^{-3}$ mol/L的HCl溶液注入石灰石$CaCO_3$层中，使之达到平衡。给出发生的化学反应，并计算平衡时水的硬度和碱度，对$1.00×10^{-3}$ mol/L HF的废液也做同样的计算。

3-6 对于在$[H^+]=4.69×10^{-11}$时，总碱度为$1.00×10^{-3}$ mol/L的溶液（来自HCO_3^-、CO_3^{2-}和OH^-的贡

---

❶ 1pound（磅）= 453.59237g。

alkalinity from CO_3^{2-}?

3-7 A wastewater disposal well for carrying various wastes at different times is drilled into a formation of limestone ($CaCO_3$), and the wastewater has time to come to complete equilibrium with the calcium carbonate before leaving the formation through an underground aquifer. Of the following components in the wastewater, the one that would not cause an increase in alkalinity due either to the component itself or to its reaction with limestone, is: (1) NaOH, (2) CO_2, (3) HF, (4) HCl, (5) all of the preceding would cause an increase in alkalinity.

3-8 Calculate the ratio $[PbT^-]/[HT^{2-}]$ for NTA in equilibrium with $PbCO_3$ in a medium having $[HCO_3^-] = 3.00 \times 10^{-3}$ mol/L.

3-9 If the medium in Problem 3-8 contained excess calcium such that the concentration of uncomplexed calcium, $[Ca^{2+}]$, were 5.00×10^{-3} M, what would be the ratio $[PbT^-]/[CaT^-]$ at pH=7?

3-10 A wastewater stream containing 1.00×10^{-3} M disodium NTA, Na_2HT, as the only solute is injected into a limestone ($CaCO_3$) formation through a waste disposal well. After going through this aquifer for some distance and reaching equilibrium, the water is sampled through a sampling well. What is the

3-7 一口用于输送不同时期各种废物的废水处理井钻入石灰石（$CaCO_3$）的地层中，废水有时间与碳酸钙完全平衡，然后通过地下含水层离开地层。废水中下列组分中，不会因组分本身或与石灰石反应而引起碱度升高的是：（1）NaOH；（2）CO_2；（3）HF；（4）HCl；（5）前面的都会引起碱度升高。

3-8 在具有$[HCO_3^-] = 3.00 \times 10^{-3}$mol/L的介质中，NTA与$PbCO_3$平衡。计算$[PbT^-]/[HT^{2-}]$。

3-9 如果问题3-8中的培养基中含有过量的钙，使未配合的钙$[Ca^{2+}]$的浓度为5.00×10^{-3}M，那么在pH=7时，$[PbT^-]/[CaT^-]$的比值是多少？

3-10 含有1.00×10^{-3}M的NTA二钠（Na_2HT）作为唯一的溶质的废水，通过一个废物处理井注入石灰石（$CaCO_3$）层。在通过该含水层一段距离并达到平衡后，通过取样井对水进行取样，NTA和$CaCO_3$

reaction between NTA species and $CaCO_3$? What is the equilibrium constant for the reaction? What are the equilibrium concentrations of CaT^-, HCO_3^-, and HT^{2-}? (The appropriate constants may be looked up in this chapter)

3-11 If the wastewater stream in Problem 3-10 were 0.100mol/L in NTA and contained other solutes that exerted a buffering action such that the final pH were 9.00, what would be the equilibrium value of HT^{2-} concentration in moles/liter?

3-12 Exactly 1.00×10^{-3} mole of $CaCl_2$, 0.100 mole of NaOH, and 0.100 mole of Na_3T were mixed and diluted to 1.00 liter. What was the concentration of Ca^{2+} in the resulting mixture?

3-13 How does chelation influence corrosion?

3-14 The following ligand has more than one site for binding to a metal ion. How many such sites does it have?

$$^-O-\overset{O}{\underset{\|}{C}}-\overset{H}{\underset{H}{C}}-\overset{H}{\underset{|}{N}}-\overset{H}{\underset{H}{C}}-\overset{O}{\underset{\|}{C}}-O^-$$

3-15 If a solution containing initially 25mg/L trisodium NTA is allowed to come to equilibrium with solid $PbCO_3$ at pH=8.50 in a medium that contains 1.76×10^{-3} mol/L HCO_3^- at equilibrium, what is the value of the ratio of the concentration of NTA bound with lead to the concentration of unbound NTA, $[PbT^-]/[HT^{2-}]$?

3-11 如果问题 3-10 中的废水中的 NTA 含量为 0.100mol/L,并含有其他溶质发挥缓冲作用,使最终 pH 值为 9.00,则 HT^{2-} 的平衡浓度为多少 mol/L?

3-12 将正好 1.00×10^{-3} mol 的 $CaCl_2$、0.100mol 的 NaOH 和 0.100mol 的 Na_3T 混合并稀释到 1.00L。所得混合物中 Ca^{2+} 的浓度是多少?

3-13 螯合是如何影响腐蚀的?

3-14 以下配体有不止一个金属离子结合位点,它有多少个这样的位点?

3-15 如果在含有 1.76×10^{-3} mol/L 的 HCO_3^- 的介质中,让最初含有 25mg/L 三钠 NTA 的溶液在 pH=8.50 时与固体 $PbCO_3$ 达到平衡,则与铅结合的 NTA 浓度与未结合的 NTA 浓度之比 $[PbT^-]/[HT^{2-}]$ 的值是多少?

3-16 After a low concentration of NTA has equilibrated with $PbCO_3$ at pH = 7.00 in a medium having $[HCO_3^-] = 7.50 \times 10^{-4}$ mol/L, what is the ratio of $[PbT^-]/[HT^{2-}]$?

3-17 What detrimental effect may dissolved chelating agents have upon conventional biological waste treatment?

3-18 Why is chelating agent usually added to artificial algal growth media?

3-19 What common complex compound of magnesium is essential to certain life processes?

3-20 What is always the ultimate product of polyphosphate hydrolysis?

3-21 A solution containing initially 1.00×10^{-5} mol/L CaT^- is brought to equilibrium with solid $PbCO_3$. At equilibrium, pH = 7.00, $[Ca^{2+}] = 1.50 \times 10^{-3}$ mol/L, and $[HCO_3^-] = 1.10 \times 10^{-3}$ mol/L. At equilibrium, what is the fraction of total NTA in solution as PbT^-?

3-22 What is the fraction of NTA present as HT^{2-} after HT^{2-} has been brought to equilibrium with solid $PbCO_3$ at pH = 7.00 in a medium in which $[HCO_3^-] = 1.25 \times 10^{-3}$ mol/L.

3-23 Describe ways in which measures taken to alleviate water supply and flooding problems might actually aggravate such problems.

3-16 低浓度的 NTA 与 $PbCO_3$ 在 pH=7.00 的介质中平衡后，$[HCO_3^-]=7.50\times10^{-4}$ mol/L，$[PbT^-]/[HT^{2-}]$ 的比例是多少?

3-17 溶解的螯合剂会对传统的生物污水处理产生哪些不利影响?

3-18 为什么人工藻类生长培养基中通常要添加螯合剂?

3-19 镁的什么常见复合化合物是某些生命过程中必不可少?

3-20 多磷酸盐水解的最终产物是什么?

3-21 将最初含有 1.00×10^{-5} mol/L CaT^- 的溶液与固体 $PbCO_3$ 达到平衡。在平衡时，pH=7.00，$[Ca^{2+}]=1.50\times10^{-3}$ mol/L，$[HCO_3^-]=1.10\times10^{-3}$ mol/L。在平衡状态下，PbT^- 占溶液中 NTA 总量的比例是多少?

3-22 在 $[HCO_3^-]=1.25\times10^{-3}$ mol/L 的介质中，在 HT^{2-} 与 pH=7.00 的固体 $PbCO_3$ 达到平衡后，作为 HT^{2-} 的 NTA 的含量是多少。

3-23 哪些为缓解供水和水灾问题而采取的措施，实际上可能会加剧这些问题?

3-24 The study of water is known as _____, _____ is the branch of the science dealing with the characteristics of fresh water, and the science that deals with about 97% of all Earth's water is called _____.

3-25 Consider the unique and important properties of water. What molecular or bonding characteristics of the water molecules are largely responsible for these properties. List or describe one of each of the following unique properties of water related to.

(1) Thermal characteristics.

(2) Transmission of light.

(3) Surface tension.

(4) Solvent properties.

3-26 Relate aquatic life to aquatic chemistry. In so doing, consider the following：

(1) autotrophic organisms, (2) producers,

(3) heterotrophic organisms, (4) decomposers,

(5) eutrophication, (6) dissolved oxygen,

(7) biochemical oxygen demand.

3-27 Assuming levels of atmospheric CO_2 are 4.0×10^{-4} CO_2, what is the pH of rainwater due to the presence of carbon dioxide? Some estimates are for atmospheric carbon dioxide levels to double in the future. What would be the pH of rainwater if this happens?

3-24 关于水的研究被称为_____，_____是处理淡水特性的科学分支，而处理地球上约97%的水的科学被称为_____。

3-25 考虑水的独特而重要的性质，这些性质主要由水分子的哪些分子或键合特性决定？列举或描述水的下列独特性质。

(1) 热特性；

(2) 透光性；

(3) 表面张力；

(4) 溶剂特性。

3-26 将水生生物与水化学联系起来。在此过程中，考虑以下内容：

(1) 自养生物；(2) 生产者；(3) 异养生物；(4) 分解者；(5) 富营养化；(6) 溶解氧；(7) 生化需氧量。

3-27 假设大气中CO_2的含量为4.0×10^{-4}，由于CO_2的存在使得雨水的pH值是多少？有人估计大气中的二氧化碳水平在未来会翻倍。如果发生这种情况，雨水的pH值会是多少？

3-28 Assume a sewage treatment plant processing 1 million liters of wastewater per day containing 200mg/L of degradable biomass, {CH$_2$O}. Calculate the volume of dry air at 25℃ that must be pumped into the wastewater per day to provide the oxygen required to degrade the biomass.

3-29 Anoxic bacteria growing in a lake sediment produced equal molar amounts of carbon dioxide and carbon monoxide according to the biochemical reaction 2{CH$_2$O}→CO$_2$+CH$_4$, so that the water in the lake was saturated with both CO$_2$ gas and CH$_4$ gas. In units of mol/(L·atm)$^{-1}$ the Henry's law constant for CO$_2$ is 3.38×10^{-2} and that of CH$_4$ has a value of 1.34×10^{-3}. At the depth at which the gas was being evolved, the total pressure was 1.10atm❶ and the temperature was 25℃, so the vapor pressure of water was 0.0313atm❶. Calculate the concentrations of dissolved CO$_2$ and dissolved CH$_4$.

3-30 The acid-base reaction for the dissociation of acetic acid is: HOAc+H$_2$O→H$_3$O$^+$+OAc$^-$ with $K_a = 1.75\times10^{-5}$. Break this reaction down into two half-reactions involving H$^+$ ion. Break down the redox reaction Fe^{2+}+H$^+$→Fe^{3+}+1/2H$_2$ into two half-reactions involving the electron. Discuss the analogies between the acid-base and redox processes.

3-28 假设一家污水处理厂每天处理 1×10^6L 废水，其中含有 200mg/L 可降解生物质 {CH$_2$O}。计算每天必须泵入废水的 25℃ 的干燥空气量，以提供降解生物质所需的氧气。

3-29 在湖泊沉积物中生长的缺氧菌根据生化反应 {CH$_2$O}→CO$_2$+CH$_4$ 产生等摩尔量的二氧化碳和甲烷，使湖中的水同时饱和了 CO$_2$ 气体和 CH$_4$ 气体。单位为 mol/(L·atm)$^{-1}$，CO$_2$ 的亨利定律常数为 3.38×10^{-2}，CH$_4$ 的亨利定律常数为 1.34×10^{-3}。在气体演化的深度上，总压强为 1.10atm❶，温度为 25℃，所以水的蒸气压为 0.0313atm❶。计算溶解的 CO$_2$ 和溶解的 CH$_4$ 的浓度。

3-30 乙酸解离的酸碱反应为 HOAc + H$_2$O → H$_3$O$^+$ + OAc$^-$，$K_a = 1.75\times10^{-5}$。将此反应分解为两个涉及 H$^+$ 离子的半反应。将氧化还原反应 Fe^{2+}+H$^+$→Fe^{3+}+1/2H$_2$ 分解成涉及电子的两个半反应。讨论酸酸反应和氧化还原之间的类比。

---

❶ 1atm = 101.325kPa。

3-31 Assuming a bicarbonate ion concentration $[HCO_3^-]$ of 1.00×10^{-3} M and a value of 3.5×10^{-11} for the solubility product of $FeCO_3$, what would you expect to be the stable iron species at pH = 9.5 and pE = -8.0?

3-32 Assuming that the partial pressure of oxygen in water is that of atmospheric O_2, 0.21atm, rather than the 1.00atm❶, derive an equation describing the oxidizing pE limit of water as a function of pH.

3-33 Plot $\lg P_{O_2}$ as a function of pE at pH = 7.00.

3-34 Calculate the values of $[Fe^{3+}]$, pE, and pH at the point where Fe^{2+} at a concentration of 1.00×10^{-5} M, Fe$(OH)_2$, and Fe$(OH)_3$ are all in equilibrium.

3-35 What is the pE value in a solution in equilibrium with air (21% O_2 by volume) at pH = 6.00?

3-36 What is the pE value at the point on the Fe^{2+}-Fe$(OH)_3$ boundary line in a solution with a soluble iron concentration of 1.00×10^{-4} M at pH = 6.00?

3-37 What is the pE value in an acid mine water sample having $[Fe^{3+}] = 7.03 \times 10^{-3}$ M and $[Fe^{2+}] = 3.71 \times 10^{-4}$ M?

3-38 At pH = 6.00 and pE = 2.58, what is the concentration of Fe^{2+} in equilibrium with Fe$(OH)_3(s)$?

3-39 What is the calculated value of the partial pressure of O_2 in acid mine water of pH =

3-31 假设碳酸氢根离子 $[HCO_3^-]$ 浓度为 1.00×10^{-3} M，$FeCO_3$ 的溶度积数值为 3.5×10^{-11}，在 pH = 9.5 和 pE = -8.0 下稳定的铁的种类是什么？

3-32 假设水中的氧分压是大气中 O_2 的分压，即 0.21atm，而不是 1.00atm❶，推导出描述水的氧化 pE 极限与 pH 值函数关系的公式。

3-33 绘制 pH 值为 7.00 时 P_{O_2} 作为 pE 的函数。

3-34 Fe^{2+} 浓度为 1.00×10^{-5} M，Fe$(OH)_2$ 和 Fe$(OH)_3$ 均处于平衡，计算 $[Fe^{3+}]$、pE 和 pH 值在该点的值。

3-35 在 pH = 6.00 时，与空气（O_2 体积分数为 21%）平衡的溶液的 pE 值是多少？

3-36 在 pH = 6.00 的可溶性铁浓度为 1×10^{-4} mol/L 的溶液中，Fe^{2+}-Fe$(OH)_3$ 边界线上的点的 pE 值是多少？

3-37 酸性矿井水样中 $[Fe^{3+}] = 7.03 \times 10^{-3}$ M，$[Fe^{2+}] = 3.71 \times 10^{-4}$ M，pE 是多少？

3-38 在 pH = 6.00 和 pE = 2.58 时，与 Fe$(OH)_3(s)$ 平衡的 Fe^{2+} 浓度是多少？

3-39 在 pH = 2.00 的酸性矿井水中，$[Fe^{3+}] = [Fe^{2+}]$，O_2

---

❶ 1atm = 101.325kPa。

2.00, in which $[Fe^{3+}] = [Fe^{2+}]$?

3-40 What is the major advantage of expressing redox reactions and half-reactions in terms of exactly one electron-mole?

3-41 Why are pE values that are determined by reading the potential of a platinum electrode versus a reference electrode generally not very meaningful?

3-42 What determines the oxidizing and reducing limits, respectively, for the thermodynamic stability of water?

3-43 How would you expect pE to vary with depth in a stratified lake?

3-44 Upon what half-reaction is the rigorous definition of pE based?

3-45 Analysis of water in a sediment sample at equilibrium at pH=7.00 showed $[SO_4^{2-}] = 2.00 \times 10^{-5}$ M and a partial pressure of H_2S of 0.100 atm❶. Show with appropriate calculations if methane, CH_4, would be expected in the sediment.

3-46 Choose the correct answer of the following, explain why it is true, and explain why the other choices are untrue.

A. High pE is associated with species such as CH_4, NH_4^+, and Fe^{2+}

B. Low pE is associated with species such as CO_2, O_2, and NO_3^-

C. Values of pE in bodies of water range from about 1×10^{-7} to about 1×10^7

D. pE is a number, but cannot be related to anything real, such as is the case with pH

E. pE uses convenient numbers to express

3-40 用正好一个电子摩尔来表示氧化还原反应和半反应的主要优点是什么?

3-41 为什么通过读取铂电极相比参比电极的电位确定的 pE 值一般意义不大?

3-42 水的热力学稳定性的氧化极限和还原极限分别由什么决定?

3-43 在分层湖中，pE 随深度如何变化?

3-44 pE 的严格定义是基于什么半反应?

3-45 在 pH=7.00 的平衡状态下，对沉积物样品中的水进行分析，结果显示 $[SO_4^{2-}] = 2.00 \times 10^{-5}$ M，H_2S 的分压为 0.100 atm❶。用适当的计算显示是否会在沉积物中产生甲烷 CH_4。

3-46 从以下选项中选择正确答案，解释其正确性，并解释其他选择为何不正确。

A. 与高 pE 相关的物种，例如 CH_4、NH_4^+、Fe^{2+}

B. 与低 pE 相关的物种，例如 CO_2、O_2、NO_3^-

C. 水体中 pE 值范围为 1×10^{-7} 至 1×10^7

D. pE 是一个数字，但不能与任何真实的东西（比如 pH 值）相关

E. pE 使用方便的数字表

---

❶ 1 atm = 101.325 kPa。

electron activity over many orders of magnitude

3-47 Match each of the following from the lettered list to the reaction that corresponds to it from the numbered list below:

(A) For 1 electron-mole

(B) Reaction for standard electrode

(C) At upper pE limit of water

(D) Formation of a pollutant when anoxic water is brought to the surface

(1) $Fe(H_2O)_6^{2+} \rightleftharpoons e^- + Fe(OH)_3(s) + 3H_2O + 3H^+$

(2) $H_2 \rightleftharpoons 2H^+ + 2e^-$

(3) $1/8 NH_4^+ + 1/4 O_2 \rightleftharpoons 1/8 NO_3^- + 1/4 H^+ + 1/8 H_2O$

(4) $H_2O \rightleftharpoons O_2 + 4H^+ + 4e^-$

3-48 Of the following, the true statement regarding oxidation-reduction reactions and phenomena in natural water systems is (   ).

A. at a pE higher than the oxidizing limit of stability water decomposes to evolve H_2

B. the production of CH_4 at a very low pE is caused to occur by the action of bacteria

C. in the pE–pH diagram for iron, the region of greatest area is occupied by solid $Fe(OH)_2$

D. it is easy to accurately measure pE of water with a platinum electrode

E. there are no pE/pH limits for the regions of stability of H_2O

3-49 A sediment sample was taken from a lignite strip-mine pit containing highly alkaline (pH=10) water. Cations were

3-47 将以下字母列表中的各项与对应的编号列表的反应相匹配：

(A) 1 电子摩尔

(B) 标准电极的反应

(C) 水的 pE 上限

(D) 将缺氧水带到地表时形成污染物

(1) $Fe(H_2O)_6^{2+} \rightleftharpoons e^- + Fe(OH)_3(s) + 3H_2O + 3H^+$

(2) $H_2 \rightleftharpoons 2H^+ + 2e^-$

(3) $1/8 NH_4^+ + 1/4 O_2 \rightleftharpoons 1/8 NO_3^- + 1/4 H^+ + 1/8 H_2O$

(4) $H_2O \rightleftharpoons O_2 + 4H^+ + 4e^-$

3-48 下列关于自然水系中的氧化-还原反应和现象，说法正确的是（   ）。

A. 在 pE 高于稳定性的氧化极限时，水会分解生成 H_2

B. 在极低的 pE 下，CH_4 的生成是由细菌的作用引起的

C. 在铁的 pE–pH 图中，面积最大的区域被固体 $Fe(OH)_2$ 占据

D. 用铂电极很容易准确测量水的 pE

E. H_2O 的稳定性区域没有 pE/pH 限制

3-49 从含有高碱性水（pH=10）的褐铁矿带矿井中采集沉积物样品。用 HCl

displaced from the sediment by treatment with HCl. A total analysis of cations in the leachate yielded, on the basis of millimoles per 100g of dry sediment, 150mmol of Na^+, 5mmol of K^+, 20mmol of Mg^{2+}, and 75mmol of Ca^{2+}. What is the cation exchange capacity of the sediment in milliequivalents per 100g of dry sediment? Why does H^+ not have to be considered in this case?

3-50 What is the value of $[O_2(aq)]$ for water saturated with a mixture of 50% O_2, 50% N_2 by volume at 25℃ and a total pressure of 1.00atm❶?

3-51 Of the following, the least likely mode of transport of iron(Ⅲ) in a normal stream is ( ).
A. bound to suspended humic material
B. bound to clay particles by cation exchange processes
C. as suspended Fe_2O_3
D. as soluble Fe^{3+} ion
E. bound to colloidal clay-humic substance complexes

3-52 How does freshly precipitated colloidal iron(Ⅲ) hydroxide interact with many divalent metal ions in solution?

3-53 What stabilizes colloids composed of bacterial cells in water?

3-54 The solubility of oxygen in water is 14.74mg/L at 0℃ and 7.03mg/L at

---

❶ 1atm = 101.325kPa。

35℃. Estimate the solubility at 50℃.

3-55 What is thought to be the mechanism by which bacterial cells aggregate?

3-56 What is a good method for the production of freshly precipitated MnO_2?

3-57 A sediment sample was equilibrated with a solution of NH_4^+ ion, and the NH_4^+ was later displaced by Na^+ for analysis. A total of 33.8 milliequivalents of NH_4^+ were bound to the sediment and later displaced by Na^+. After drying, the sediment weighed 87.2g. What was its CEC in mmol/g?

3-58 A sediment sample with a CEC of 67.4 milliequivalents/g was found to contain the following exchangeable cations in milliequivalents/g: Ca^{2+}, 21.3×10^3; Mg^{2+}, 5.2×10^2; Na^+, 4.4×10^2; K^+, 0.7×10^2. The quantity of hydrogen ion, H^+, was not measured directly. What was the ECS of H^+ in milliequivalents/g?

3-59 What is the meaning of zero point of charge as applied to colloids? Is the surface of a colloidal particle totally without charged groups at the ZPC?

3-60 The concentration of methane in an interstitial water sample was found to be 150 mL/L at STP. Assuming that the methane was produced by the fermentation of organic matter, {CH_2O}, what mass of organic matter was required to produce the methane in a liter of the interstitial water?

3-61 What is the difference between CEC and ECS?

3-62 Match the sedimentary mineral from the top list designated with letters with its conditions of formation from the bottom list designated with numbers:

(A) FeS(s)

(B) $Ca_5OH(PO_4)_3$

(C) $Fe(OH)_3$

(D) $CaCO_3$

(1) May be formed when anoxic water is exposed to O_2.

(2) May be formed when oxic water becomes anoxic.

(3) Photosynthesis byproduct.

(4) May be formed when wastewater containing a particular kind of contaminant flows into a body of very hard water.

3-63 In terms of their potential for reactions with species in solution, how might metal atoms, M, on the surface of a metal oxide, MO, be described?

3-64 Air is 20.95% oxygen by volume. If air at 1.00atm❶ pressure is bubbled through water at 25℃, what is the partial pressure of O_2 in the water?

3-65 The volume percentage of CO_2 in a mixture of that gas with N_2 was determined by bubbling the mixture at 1.00atm❶ and 25℃ through a solution of 0.0100M $NaHCO_3$ and measuring the pH. If the equilibrium pH was 6.50, what was the volume percentage of CO_2?

3-62 将字母列表中的沉积矿物与数字列表中的沉积条件进行匹配:

(A) FeS(s)

(B) $Ca_5OH(PO_4)_3$

(C) $Fe(OH)_3$

(D) $CaCO_3$

(1) 可能是缺氧水接触O_2时形成的。

(2) 可能是氧化水变成缺氧水时形成的。

(3) 光合作用的副产物。

(4) 含有某种污染物的废水流入极硬水体时可能形成。

3-63 从与溶液中的物种发生反应的可能性来看,如何描述金属氧化物 MO 表面的金属原子 M?

3-64 空气中氧气的体积分数为20.95%。如果在25℃时,用1.00atm❶的空气向水中鼓泡,则水中O_2的分压是多少?

3-65 测定N_2和CO_2混合气体中CO_2的体积百分比,是在1.00atm❶和25℃下将混合气体通入0.0100M的$NaHCO_3$溶液进行鼓泡,并测量 pH 值。若平衡时 pH 为6.50,则CO_2的体积分数是多少?

---

❶ 1atm=101.325kPa。

3-66 For what purpose is a polymer with the following general formula used?

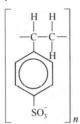

3-66 具有以下通式的聚合物有什么用途?

3-67 Of the following statements, the one that is true regarding colloids is (   ).
A. hydrophilic colloids consist of aggregates of relatively small molecules
B. hydrophobic colloids do not have electrical charges
C. hydrophilic colloids are those formed by clusters of species, such as $H_3C(CH_2)_{16}CO_2^-$
D. association colloids form micelles
E. the electrical charges of hydrophobic colloids are insignificant

3-67 下列关于胶体的说法,正确的是(   )。
A. 亲水性胶体是由相对较小的分子聚集而成的
B. 疏水性胶体不带电荷
C. 亲水胶体是由物种簇形成,如 $H_3C(CH_2)_{16}CO_2^-$
D. 缔合胶体形成胶束
E. 疏水性胶体的电荷是不重要的

3-68 For a slightly soluble divalent metal sulfate, MSO_4, $K_{sp} = 9.00 \times 10^{-14}$. An excess of pure solid MSO_4 was equilibrated with pure water to give a solution which contains 6.45×10^{-7} mol/L of dissolved M. Considering these observations the true statement is (   ).

A. MSO_4 has a significant degree of intrinsic solubility
B. the solubility product, alone, accurately predicts solubility
C. the value of the solubility product is in error
D. the concentration of M in water must

3-68 对于微溶的二价金属硫酸盐 MSO_4,$K_{sp} = 9.00 \times 10^{-14}$。将过量的纯固体 MSO_4 与纯水平衡,得到的溶液中含有 6.45×10^{-7} mol/L 溶解的 M,根据这些观察,下列说法正确的是(   )。
A. MSO_4 具有相当程度的固有溶解度
B. 仅凭溶度积就能准确预测溶解度
C. 溶度积的数值有误
D. 水中 M 的浓度一定

have been in error

E. the only explanation for the observations is formation of HSO_4^-

3-69 Of the following, the incorrect statement regarding sediments and their formation is (   ).

A. physical, chemical, and biological processes may all result in the deposition of sediments in the bottom regions of bodies of water

B. indirectly, photosynthesis can result in formation of $CaCO_3$ sediment

C. oxidation of Fe^{2+} ion can result in formation of an insoluble species that can be incorporated into sediment

D. sediments typically consist of mixtures of clay, silt, sand, organic matter, and various minerals, and may vary in composition from pure mineral matter to predominantly organic matter

E. FeS that gets into sediment tends to form at the surface of water in contact with O_2

3-70 Given that at 25℃, the Henry's law constant for oxygen is $1.28×10^{-3} mol/(L \cdot atm)^{-1}$ and the partial pressure of water vapor is 0.0313atm❶, what is the value of $[O_2(aq)]$ for water saturated with a mixture of 33.3% O_2, 66.7% N_2 by volume at 25℃ and a total pressure of 1.00 atm in units of mol/L?

3-69 下列关于沉积物及其形成的说法，不正确的是（   ）。

A. 物理、化学、生物过程都可能导致沉积物在水体底部区域的沉积

B. 光合作用可间接导致 $CaCO_3$ 沉积物的形成

C. Fe^{2+} 离子的氧化作用可形成不溶性物种，可融入沉积物中

D. 沉积物通常由黏土、淤泥、沙子、有机物和各种矿物质的混合物组成，其成分可能从纯矿物物质到以有机物为主

E. 进入沉积物的 FeS 往往在水面与 O_2 接触后形成

3-70 设 25℃ 时，氧气的亨利定律常数为 $1.28×10^{-3} mol/(L \cdot atm)^{-1}$，水蒸气分压为 0.0313atm❶，则在 25℃ 下，总压力为 1.00atm、体积分数 33.3% O_2、66.7% N_2 的混合物达到饱和的水中，$[O_2(aq)]$ 的值是多少（单位为 mol/L）？

---

❶ 1atm=101.325kPa。

3-71 Match the following regarding colloids from the top list designated with letters with its corresponding match from the bottom list designated with numbers.

(A) Hydrophilic colloids

(B) Association colloids

(C) Hydrophobic colloids

(D) Noncolloidal

(1) $CH_3CO_2^-Na^+$

(2) Macromolecular proteins

(3) Often removed by addition of salt

(4) $CH_3(CH_2)_{16}CO_2^-Na^+$

3-72 As $CH_3CH_2CH_2CH_2CO_2H$ biodegrades in several steps to carbon dioxide and water, various chemical species are observed. What stable chemical species would be observed as a result of the first step of this degradation process?

3-73 ( ) is true regarding the production of methane in water.

A. it occurs in the presence of oxygen

B. it consumes oxygen

C. it removes biological oxygen demand from the water

D. it is accomplished by oxic bacteria

E. it produces more energy per electron-mole than does oxic respiration

3-74 Suppose that the anoxic fermentation of organic matter, {CH_2O}, in water

3-71 将下列用字母表示的胶体与用数字表示的论述进行匹配。

(A) 亲水胶体

(B) 缔合胶体

(C) 疏水胶体

(D) 非胶体

(1) $CH_3CO_2^-Na^+$

(2) 大分子蛋白质

(3) 常因加盐而去除

(4) $CH_3(CH_2)_{16}CO_2^-Na^+$

3-72 $CH_3CH_2CH_2CH_2CO_2H$ 经过几个步骤生物降解为二氧化碳和水时,会观察到各种化学物种。这一降解过程的第一步会观察到什么稳定的化学物种?

3-73 关于水中甲烷的产生,( ) 是正确的。

A. 在氧气存在的情况下发生

B. 消耗氧气

C. 消除了水中的生物需氧量

D. 由氧化细菌完成

E. 每电子摩尔产生的能量比氧化呼吸多

3-74 假设有机物 {CH_2O} 在水中的缺氧发酵产生

yields 15.0L of CH_4 (at standard temperature and pressure). How many grams of oxygen would be consumed by the oxic respiration of the same quantity of {CH_2O}?

15.0L CH_4（在标准温度和压力下），同量的{CH_2O}进行氧化呼吸要消耗多少克氧气？

3-75 Most plants assimilate nitrogen as nitrate ion. However, ammonia (NH_3) is a popular and economical fertilizer. What essential role do bacteria play when ammonia is used as a fertilizer? Do you think any problems might occur when using ammonia in a waterlogged soil lacking oxygen?

3-75 大多数植物将氮同化为硝酸根离子，然而氨（NH_3）是一种受欢迎的经济型肥料。当氨作为肥料使用时，细菌起着什么重要作用，你认为在缺氧的积水土壤中使用氨会出现什么问题吗？

3-76 Wastewater containing 8mg/L O_2 (atomic mass O = 16), 1.00×10^{-3} mol/L NO_3^-, and 1.00×10^{-2} mol/L soluble organic matter, {CH_2O}, is stored isolated from the atmosphere in a container richly seeded with a variety of bacteria. Assume that denitrification is one of the processes which will occur during storage. After the bacteria have had a chance to do their work, (　　) will be true?

A. no {CH_2O} will remain

B. some O_2 will remain

C. some NO_3^- will remain

D. denitrification will have consumed more of the organic matter than oxic respiration

E. the composition of the water will remain unchanged

3-76 含有 8mg/L O_2（原子质量 O = 16）、1.00×10^{-3} mol/L NO_3^- 和 1.00×10^{-2} mol/L 的可溶性有机物 {CH_2O} 的废水，储存在一个与大气隔离的容器中，容器中富含各种细菌。假设反硝化作用是储存过程中会发生的反应之一，在细菌有机会作用后，（　　）是正确的。

A. 不会有 {CH_2O} 残留

B. 会残留一些 O_2

C. 会残留一些 NO_3^-

D. 反硝化会比氧化呼吸消耗掉更多的有机物

E. 水的组分将保持不变

3-77 What factors favor the production of methane in anoxic surroundings?

3-78 (   ) is true regarding chromium in water.
A. chromium(Ⅲ) is suspected of being carcinogenic
B. chromium(Ⅲ) is less likely to be found in a soluble form than chromium(Ⅵ)
C. the toxicity of chromium(Ⅲ) in electroplating wastewaters is decreased by oxidation to chromium(Ⅵ)
D. chromium is not an essential trace element
E. chromium is known to form methylated species analogous to methylmercury compounds

3-79 What do mercury and arsenic have in common in regard to their interactions with bacteria in sediments?

3-80 Of the compounds $CH_3(CH_2)_{10}CO_2H$, $(CH_3)_3C(CH_3)_2CO_2H$, $CH_3(CH_2)_{10}CH_3$, and $\phi-(CH_2)_{10}CH_3$ (where ϕ represents a benzene ring), which is the most readily biodegradable?

3-81 Of the following, the one that is not a cause of, or associated with eutrophication is (   ).
A. eventual depletion of oxygen in the water
B. excessive phosphate
C. excessive algal growth

3-77 哪些因素有利于缺氧环境中甲烷的产生？

3-78 关于水中的铬，（   ）是正确的。
A. 铬（Ⅲ）具有疑似致癌性
B. 铬（Ⅲ）以可溶性形式存在的可能性比铬（Ⅵ）小
C. 电镀废水中铬（Ⅲ）的毒性会因氧化成铬（Ⅵ）而降低
D. 铬不是必需的微量元素
E. 已知铬会形成类似甲基汞化合物的甲基化物种

3-79 汞和砷在与沉积物中的与细菌的相互作用方面有什么共同之处？

3-80 在化合物 $CH_3(CH_2)_{10}CO_2$，$(CH_3)_3C(CH_3)_2CO_2H$、$CH_3(CH_2)_{10}CH_3$、$\phi-(CH_2)_{10}CH_3$（其中 ϕ 代表苯环）中，哪种化合物最容易生物降解？

3-81 下列各项中，不属于富营养化原因或不与富营养化有关的是（   ）。
A. 最终耗尽水中的氧气
B. 磷酸盐过多
C. 藻类生长过多

D. excessive nutrients

E. excessive O_2

3-82 Of the following heavy metals, choose the one most likely to have microorganisms involved in its mobilization in water and explain why this is so. (　　)

A. lead

B. mercury

C. cadmium

D. chromium

E. zinc

3-83 Of the following, the true statement is (　　).

A. eutrophication results from the direct discharge of toxic pollutants into water

B. treatment of a lake with phosphates is a process used to deter eutrophication

C. alkalinity is the most frequent limiting nutrient in eutrophication

D. eutrophication results from excessive plant or algal growth

E. eutrophication is generally a beneficial phenomenon because it produces oxygen

3-84 How many moles of NTA should be added to 1000L of water having a pH of 9 and containing CO_3^{2-} at 1.00×10^{-4} M to prevent precipitation of $CaCO_3$? Assume a total calcium level of 40mg/L.

3-85 How many liters of methanol would be required daily to remove the nitrogen from a 200000-L/day sewage treatment plant producing an effluent containing 50mg/L of nitrogen? Assume that the

D. 营养物质过多

E. O_2 过多

3-82 在下列重金属中，请选择最可能有微生物参与其在水中的迁移的一种，并解释为什么会这样。(　　)

A. 铅

B. 汞

C. 镉

D. 铬

E. 锌

3-83 下列说法中正确的是(　　)。

A. 富营养化是有毒污染物直接排入水中造成的

B. 用磷酸盐处理湖泊是阻止富营养化的过程

C. 碱度是富营养化中最常见的限制性营养物质

D. 富营养化是植物或藻类过度生长造成的

E. 富营养化一般是一种有益现象，因为它能产生氧气

3-84 为了防止 $CaCO_3$ 的沉淀，应向 1000L pH 值为 9，含有 CO_3^{2-} 的水中加入多少摩尔的 NTA？假设总的钙含量为 40mg/L。

3-85 一个日处理量 200000L 的污水处理厂，其出水含氮量为 50mg/L，每天需要多少升甲醇用于脱氮？假设氮在该厂已转化为

nitrogen has been converted to NO_3^- in the plant. The denitrifying reaction is:

$$6NO_3^- + 5CH_3OH + 6H^+ \longrightarrow 3N_2(g) + 5CO_2 + 13H_2O$$

3-86 A wastewater containing dissolved Cu^{2+} ion is to be treated to remove copper. Which of the following processes would not remove copper in an insoluble form:
(1) lime precipitation, (2) cementation, (3) treatment with NTA, (4) ion exchange, (5) reaction with metallic Fe.

NO_3^-。反硝化反应是:

3-86 含有溶解的 Cu^{2+} 离子的废水要进行处理以去除铜。以下哪种工艺不能去除不溶性的铜:
（1）石灰沉淀；（2）固结；（3）用 NTA 处理；（4）离子交换；（5）与金属铁反应。

扫描二维码查看
习题答案

4 Environmental Chemistry of the Geosphere
地圈环境化学

4.1 Geosphere and Geochemistry
地圈和地球化学

4.1.1 The Fragile Solid Earth and its Relationship with the Other Environmental Spheres
脆弱的固体地球及其与其他环境圈层的关系

知识框架

Interface between geosphere and atmosphere
（1）Surface albedo affecting conditions in lower atmosphere.
（2）Exposure of surface rock from melting Arctic ice can increase temperatures.

Two atmospheric phenomena that affect geosphere
（1）Excess carbon dioxide causing warming and drought.
（2）Desertification.
（3）Acid rain that promotes erosion, harms plants growing on the geosphere.

The geosphere and hydrosphere
（1）Hydrosphere rests on geosphere or below it in aquifers.
（2）Flowing water molds the geosphere.
（3）Water picks up solutes from the geosphere.

地圈与大气之间的界面
（1）地表反照率影响低层大气条件。
（2）融化的北极冰使表层岩石暴露，会升高温度。

影响地圈的两种大气现象
（1）过量的二氧化碳导致变暖和干旱。
（2）荒漠化。
（3）酸雨加剧侵蚀，损害地圈上生长的植物。

地圈和水圈
（1）水圈位于地圈上或地下的含水层中。
（2）流动的水塑造了地圈。
（3）水从地圈中吸收溶质。

(4) Water pollutants contaminate geosphere and vice versa.

The Geosphere and the Anthrosphere

(1) Most of the anthrosphere rests on geosphere.

(2) Materials used in the anthrosphere are taken from the geosphere.

(3) Geosphere receives wastes from the anthrosphere.

The Geosphere and the Biosphere

(1) Geosphere is host for most organisms.

(2) Special importance of soil for plants.

(3) Organisms strongly affected by geosphere.

1) Nutrients for organisms from geosphere.

2) Organisms influence geosphere, especially soil.

Example: Soil humic matter.

(4) 水污染物污染地圈，反之亦然。

地圈和人类活动圈

（1）大部分人类都位于地圈上。

（2）人类活动圈中使用的材料取自地圈。

（3）地圈接收来自人类活动圈的废物。

地圈和生物圈

（1）地圈是大多数生物的宿主。

（2）土壤对植物的特殊重要性。

（3）生物受地圈影响严重。

1）生物的营养来自地圈。

2）生物影响地圈，尤其是土壤。

例：土壤腐殖质。

4.1.2 The Geosphere as a Source of Natural Capital
地圈作为自然资本的来源

知识框架

The geosphere provides:

(1) Livable environment.

(2) Minerals, essential elements, and metals required for the anthrosphere.

(3) Rare earths for advanced technologies.

(4) Lithium for rapidly growing applications of lithium-ion batteries.

(5) Place to dispose wastes.

地圈提供：

（1）宜居的环境；

（2）人类活动圈所需的矿物质、基本元素和金属；

（3）稀土用于先进技术；

（4）锂，用于锂离子电池快速增长的应用；

（5）处置废物的地方。

4.1.3 Effects of Human Activities on the Geosphere
人类活动对地圈的影响

知识框架

Largely associated with resource extraction.

(1) Surface mining.
(2) Residues from mining.
(3) Removed overburden.
(4) Mine tailings.

很大程度上与资源提取有关。

(1) 表层采矿。
(2) 采矿残余。
(3) 去除覆盖层。
(4) 矿山尾矿。

4.1.4 Air Pollution and the Geosphere
空气污染与地圈

知识框架

(1) Volcanic eruptions may cause severe air pollution.
(2) Emission of subterranean gases, particularly H_2S.
(3) Windblown dust may obscure visibility, affect health.

(1) 火山喷发可能导致严重的空气污染。
(2) 排放地下气体，特别是 H_2S。
(3) 飞尘可能会遮挡能见度，影响健康。

4.1.5 Water Pollution and the Geosphere
水污染与地圈

知识框架

(1) Geosphere contaminated by water pollutants.
Example: Heavy metals deposited in soil by polluted water.
(2) Water polluted by geosphere.
Example: Arsenic in groundwater from wells in Bangladesh.

(1) 被水污染物污染的地圈。
例：被污水沉积在土壤中的重金属。
(2) 被地圈污染的水。
例：孟加拉国井中地下水中的砷。

1) Nonpoint sources of water pollution, such as atrazine herbicide washed from soil.

2) Point sources of water pollution (usually from anthrospheric sources).

Environmental Effects of Altering the Geosphere with Respect to Streams and Floodplains

(1) Levees control flooding.

Levee failure makes flooding much worse.

(2) Properly managed and vegetated floodplains.

1) Reduce flooding.

2) Increase water infiltration to aquifers.

(3) Straightening streams by channelization has caused severe flooding and erosion.

Restoration ecology has been applied to channelized streams to restore stream meanders and reduce flooding and erosion.

1) 水污染的非点源，如从土壤中洗去的阿特拉津除草剂。

2) 水污染的点源（通常来自大气层源）。

关于溪流和洪泛区改变地圈的环境影响

（1）堤防控制洪水。

堤坝故障使洪水更严重。

（2）妥善管理和种植漫滩。

1) 减少洪水。

2) 增加含水层的水渗透。

（3）通道化拉直河道导致严重的洪灾和侵蚀。

恢复生态学已应用于河道水流，以恢复河道蜿蜒并减少洪水和侵蚀。

4.2　Soil: Earth's Lifeline
　　土壤：地球的生命线

4.2.1　Have You Thanked a Clod Today?
　　　今天你感谢土块了吗?

知识框架

（1）Productive soil with suitable climate and adequate water are essential to civilization's well-being.

（2）Soil is an essential part of the environment.

（1）具有适宜气候和充足水的生产性土壤对于文明的福祉至关重要。

（2）土壤是环境的重要组成部分。

1) Very thin, on average, if Earth were a classroom globe, average thickness of soil would be that of a human cell!

2) Essential for food production.

3) Receptor for pollutants.

(3) Agriculture.

1) Production of food and biomass raw materials.

2) Tremendous potential to affect environment.

3) Crop farming.

4) Livestock farming.

(4) Pesticides used in agriculture have environmental effects.

1）非常薄，如果地球和教室用的地球仪一样大，那么土壤的平均厚度仅为人体细胞的厚度。

2）对粮食生产至关重要。

3）污染物受体。

（3）农业。

1）食品和生物质原料的生产。

2）影响环境的巨大潜力。

3）作物种植。

4）畜牧业。

（4）农业中使用的农药对环境有影响。

重要名词和术语解析

Agriculture: The production of food by growing crops and livestock, provides for the most basic of human needs.

Crop farming: Crop farming, in which plant photosynthesis is used to produce grain, fruit, and fiber.

Livestock farming: Livestock farming, in which domesticated animals are grown for meat, milk, and other animal products.

农业：通过种植农作物和牲畜来生产食物，满足人类最基本的需求。

种植业：种植业通过光合作用来生产谷物、水果和纤维。

畜牧业：饲养牲畜以获取肉、奶和其他动物产品。

4.2.2 Structure of Soil
土壤结构

知识框架

（1）Soil is a variable mixture of materials that is the final product of rock weathering.

1）Minerals.

2）Organic matter.

（1）土壤是多种多样的材料混合物，是岩石风化的最终产物。

1）矿物质。

2）有机物。

3) Water.
4) Supports plant life.

(2) Soil organic matter is partially decayed plant matter.

3) 水。
4) 植物生命的支撑。

(2) 土壤有机质是部分腐烂的植物。

重要名词和术语解析

Soil: Soil is a variable mixture of minerals, organic matter, and water capable of supporting plant life on the earth's surface.

Horizon: Typical soils exhibit distinctive layers with increasing depth. These layers are called horizons.

Topsoil: The top layer of soil, typically several inches in thickness, is known as the A horizon, or **topsoil**. The next layer is the B horizon, or **subsoil**.

Podzol: One of the more important classes of productive soils is the **podzol** type of soil formed under relatively high rainfall conditions in temperate zones of the world.

土壤：**土壤**是矿物质、有机质和水的可变混合物，能够维持地球表面的植物生命。

地层：随深度增加典型的土壤表现出独特的层次，这些层称为**地层**。

表土：通常几英寸厚的土壤顶层称为 A 层或**表土**。下一层是 B 层或**底土**。

灰壤：生产土壤中最重要的一类是**灰壤**类型的土壤，是在世界温带地区相对较高的降雨条件下形成的。

4.2.3 Composition of Soil
土壤成分

知识框架

(1) Waterlogging (saturation of air spaces in soil with water) can release phytotoxic Fe^{2+} and Mn^{2+}.

(2) Organic matter (humus) is an important soil constituent.

(3) Decay of organic matter in soil can consume and displace O_2 in soil air.

(4) Soil solution transfers solutes between soil solids and roots.

(1) 渍水（土壤中的空气充满水）会释放出具有植物毒性的 Fe^{2+} 和 Mn^{2+}。

(2) 有机质（腐殖质）是重要的土壤成分。

(3) 土壤中有机物的腐烂可以消耗和置换土壤中的氧气。

(4) 土壤溶液在土壤固体和根之间转移溶。

重要名词和术语解析

Transpiration: The water enters the atmosphere from the plant's leaves, a process called **transpiration**.

Humus: composed of a base-soluble fraction called humic and fulvic acids, and an insoluble fraction called humin, is the residue left when bacteria and fungi biodegrade plant material.

Humification: The process by which humus is formed is called **humification**.

Amphiphile: Part of each molecule of humic substance is nonpolar and hydrophobic, and part is polar and hydrophilic.
Such molecules are called **amphiphiles**, and they form micelles in which the nonpolar parts compose the inside of small colloidal particles and the polar functional groups are on the outside.

Soil solution: The **soil solution** is the aqueous portion of soil that contains dissolved matter from soil chemical and biochemical processes in soil and from exchange with the hydrosphere and biosphere.

蒸腾作用：水从植物的叶子进入大气，这个过程称为**蒸腾作用**。

腐殖质：由一种称为腐殖酸和富里酸的碱溶性成分和一种称为胡敏素的不溶性成分组成，是细菌和真菌对植物材料进行生物降解时留下的残留物。

腐殖化：腐殖质形成的过程称为**腐殖化**。

两亲物：每个腐殖质分子的一部分为非极性和疏水性，另一部分为极性和亲水性。这样的分子称为**两亲性**，它们形成胶束，其中非极性部分组成小的胶体颗粒的内部，而极性官能团在外部。

土壤溶液：**土壤溶液**是土壤的含水部分，其中包含土壤化学和生化过程以及与水圈和生物圈交换产生的溶解物质。

4.2.4 Acid-Base and Ion-Exchange Reactions in Soil 土壤中的酸碱和离子交换反应

知识框架

(1) Soils exchange cations such as H^+, NH_4^+, Ca^{2+}.

(2) Soil acts as a pH buffer.

(3) Acidification of soil by exchange of cations with H^+:

(1) 土壤交换阳离子，例如 H^+、NH_4^+、Ca^{2+}。

(2) 土壤充当pH值缓冲剂。

(3) 通过与 H^+ 交换阳离子来酸化土壤：

$$Ca^{2+}(Soil, 土壤中) + 2CO_2 + 2H_2O \longrightarrow (H^+)_2(Soil, 土壤中) +$$
$$Ca^{2+}(Root, 植物的根中) + 2HCO_3^-$$

(4) Acidification of soil by oxidation of pyrite, testing with hydrogen peroxide:

（4）通过黄铁矿的氧化来酸化土壤，用过氧化氢进行测试：

$$FeS_2 + \frac{15}{2}H_2O_2 \longrightarrow Fe^{3+} + H^+ + 2SO_4^{2-} + 7H_2O$$

(5) Phytotoxic Al^{3+} can be released from acidified soils.

（5）植物毒性的 Al^{3+} 可以从酸化土壤中释放。

(6) Adjustment of soil acidity.

1) Neutralization of acid soils with limestone:

（6）调整土壤酸度。

1) 用石灰石中和酸性土壤：

$$(H^+)_2(Soil, 土壤中) + CaCO_3 \longrightarrow Ca^{2+}(Soil, 土壤中) + CO_2 + H_2O$$

2) Reduction of excessive soil alkalinity by addition of acid-forming materials.

① Acidic Fe^{3+}.

② Acid-forming sulfur or sulfur compounds.

2) 通过添加致酸物质减少过多的土壤碱度。

① 酸性 Fe^{3+}。

② 成硫酸或硫化合物。

Ion-Exchange Equilibria in Soils

(1) Soils tend to bind with and exchange cations.

(2) For equilibria of the type, $Na^+(Soil) + K^+ \rightleftharpoons K^+(Soil) + Na^+$, an equilibrium constant expression may be written:

$$K_c = \frac{N_K [Na^+]}{N_{Na} [K^+]}$$

in which, N is the equivalent fraction of the cation on the solid, and $[Na^+]$ and $[K^+]$ are the concentrations of these ions in the soil solution.

(3) Anion exchange is less straightforward than cation exchange.

1) At low pH, exchange of anions bound ionically to + sites on solids.

土壤中的离子交换平衡

（1）土壤倾向于与阳离子结合并交换阳离子。

（2）对于这种类型的平衡，Na^+（土壤中）$+K^+ \rightleftharpoons K^+$（土壤中）$+Na^+$，平衡常数表达式为：

$$K_c = \frac{N_K [Na^+]}{N_{Na} [K^+]}$$

式中，N 是固体中阳离子的物质的量分数，$[Na^+]$ 和 $[K^+]$ 是土壤溶液中这些离子的浓度。

（3）阴离子交换不如阳离子交换直接。

1) 在低 pH 下，阴离子键与固体上的+位点离子交换。

2) At higher pH, exchange with OH⁻ groups bound to solid surface may occur:

2) 在较高的 pH 下，可能会发生与结合在固体表面的 OH⁻ 交换的情况：

$$M{-}OH + HPO_4^{2-} \longrightarrow M{-}OPO_3H_2^- + OH^-$$

4.2.5 Macronutrients in Soil
土壤中的大量养分

知识框架

Macronutrients are required by plants in relatively large quantities, such as for structural parts.

（1）Carbon, hydrogen, oxygen from air or water.

（2）Nitrogen, phosphorus, potassium, calcium, magnesium, and sulfur from soil.

（3）Nitrogen indirectly from air.

（4）Ca^{2+} usually readily available.

（5）Also from lime treatment of soil.

（6）Mg^{2+} available when bound as exchangeable ion.

（7）Sulfate assimilated as SO_4^{2-}.

植物需要大量的营养，如结构部分。

（1）空气或水中的碳、氢、氧。

（2）土壤中的氮、磷、钾、钙、镁和硫。

（3）空气中的间接氮。

（4）Ca^{2+} 通常容易获得。

（5）还来自石灰处理土壤。

（6）结合成可交换离子时可提供 Mg^{2+}。

（7）硫酸盐同化为 SO_4^{2-}。

4.2.6 Nitrogen, Phosphorus and Potassium in Soil
土壤中的氮、磷和钾

知识框架

These elements are particularly important macronutrients.

Commonly added as fertilizers

这些元素是特别重要的常量营养素。

通常作为肥料添加

Nitrogen in Soil
土壤中的氮

（1）Nitrogen commonly assimilated by plants as NO_3^-.

（1）氮通常被植物吸收为 NO_3^-。

(2) NO_3^- also a common agricultural water pollutant.

Phosphorus in Soil

Phosphorus required in only small quantities by plants.

(1) Essential in plant nucleic acids (DNA, RNA).

(2) Assimilated as orthophosphate—$H_2PO_4^-$ or HPO_4^{2-}.

Potassium in Soil

(1) Assimilated by plants as K^+.

(2) Required in relatively large quantities by plants.

(3) Commonly added as fertilizer.

4.2.7 Micronutrients in Soil
土壤中的微量营养素

知识框架

(1) Plant micronutrients required from soil are:

1) Boron.
2) Chlorine.
3) Copper.
4) Iron.
5) Manganese.
6) Zinc.
7) Molybdenum (for N fixation).

(2) Generally micronutrients are present at adequate levels, but deficiencies can occur.

(3) Deficiencies can be due to factors that prevent micronutrient uptake.

(2) NO_3^- 也是常见的农业水污染物。

土壤中的磷

植物仅需少量磷。

(1) 植物核酸（DNA, RNA）中必不可少。

(2) 同化为正磷酸盐 $H_2PO_4^-$ 或 HPO_4^{2-}。

土壤中的钾

(1) 被植物吸收为 K^+。

(2) 植物需要的量相对较大。

(3) 通常作为肥料添加。

(1) 土壤所需的植物微量营养素是：

1) 硼；
2) 氯；
3) 铜；
4) 铁；
5) 锰；
6) 锌；
7) 钼（用于固氮）。

(2) 通常微量营养素的含量足够，但可能会出现不足。

(3) 缺乏症可能是由于阻止微量营养素吸收的因素引起的。

1) pE.

2) pH.

3) Biological factors

(4) Trace metal hyperaccumulators.

Example: "Copper flower", which may reach 1.3% dry mass copper.

(5) Phytoremediation may use plants to take up pollutant metals from soil.

1) pE；

2) pH 值；

3) 生物因素。

(4) 微量金属超积累植物。

例："铜花"，铜的干质量可能达到 1.3%。

(5) 植物修复可能利用植物从土壤中吸收污染物金属。

重要名词和术语解析

Micronutrient: Boron, chlorine, copper, iron, manganese, molybdenum (for N-fixation), and zinc are considered essential plant **micronutrients**.

Hyperaccumulator: Some plants accumulate extremely high levels of specific trace metals. Those accumulating more than 1.00 mg/g of dry weight are called **hyperaccumulators**.

Phytoremediation: The hyperaccumulation of metals by some plants has led to the idea of **phytoremediation** in which plants growing on contaminated ground accumulate metals, which are then removed with the plant biomass.

微量营养素：硼、氯、铜、铁、锰、钼（用于固氮）和锌被认为是必需的植物**微量营养素**。

超级积累植物：某些植物中积累极高水平的特定微量金属。那些积累干重超过 1.00 mg/g 的植物称为**超级积累植物**。

植物修复：一些植物对金属的过度积累导致了**植物修复**的想法，在受污染的地面上生长的植物会积累金属，然后将其与植物生物质一起去除。

4.2.8 Fertilizers
肥料

知识框架

(1) Fertilizers are substances added to soil as plant nutrients.

1) Most commonly nitrogen, phosphorus, potassium.

2) Nutrient content denoted by numbers

(1) 肥料是作为植物养分添加到土壤中的物质。

1) 最常见的氮、磷、钾。

2) 以 6-12-8 等数字表示

such as 6-12-8, which would contain:

① 6% nitrogen expressed as N.

② 12% phosphorus expressed as P_2O_5.

③ 8% potassium expressed as K_2O.

(2) Organic fertilizers must decompose to release inorganic forms of nutrients.

1) N as NO_3^-.

2) P as $H_XPO_4^{X-3}$.

3) K as K^+.

4) Farm manure is only about a 0.5-0.24-0.5 fertilizer.

Nitrogen Fertilizer

(1) Atmospheric N_2 combined with H_2 over a catalyst.

1) $N_2+3H_2 \rightarrow 2NH_3$.

2) Temperature around 500℃.

3) Pressure up to 1000atm❶.

(2) Anhydrous NH_3 can be applied directly as fertilizer.

1) Held by affinity for soil moisture.

2) High N content ($w=82\%$).

(3) NH_3 in water ($w=30\%$) is a safer fertilizer to apply.

(4) Solid ammonium nitrate, NH_4NO_3, used as fertilizers.

1) Most plants require bacterial conversion of ammonium nitrogen to assimilable NO_3^-.

2) Urea is also used as N fertilizer.

$$H_2N-\underset{\underset{}{\overset{O}{\|}}}{C}-NH_2$$

Urea (尿素)

的营养成分，其中应包含：

①6%的氮以 N 表示；

②12%的磷以 P_2O_5 表示；

③8%钾以 K_2O 表示。

(2) 有机肥料必须分解以释放无机形式的养分。

1) N 为 NO_3^-；

2) P 为 $H_XPO_4^{X-3}$；

3) K 为 K^+；

4) 农场肥料的化肥含量仅为 0.5-0.24-0.5。

氮肥

(1) 催化剂上常压氮气与氢气结合。

1) $N_2+3H_2 \rightarrow 2NH_3$；

2) 温度约 500℃；

3) 压力高达 1000atm❶。

(2) 无水 NH_3 可直接用作肥料。

1) 对土壤水分的亲和力；

2) 高氮含量（$w=82\%$）。

(3) 水中 NH_3（$w=30\%$）是更安全的肥料。

(4) 固体硝酸铵 NH_4NO_3 用作肥料。

1) 大多数植物需要将铵态氮细菌转化为可吸收的 NO_3^-；

2) 尿素也用作氮肥。

---

❶ 1atm=101.325kPa。

Phosphate Fertilizer

(1) P occurs as poorly soluble fluorapatite mineral and less commonly as hydroxyapatite.

1) $Ca_5F(PO_4)_3$.
2) $Ca_5OH(PO_4)_3$.

(2) Phosphate mineral converted to more soluble superphosphates by treatment with phosphoric or sulfuric acids.

$$2Ca_5F(PO_4)_3 + 14H_3PO_4 + 10H_2O \longrightarrow 2HF(g) + 10Ca(H_2PO_4)_2 \cdot H_2O$$

$$2Ca_5F(PO_4)_3 + 7H_2SO_4 + 3H_2O \longrightarrow 2HF(g) + 3Ca(H_2PO_4)_2 \cdot H_2O + 7CaSO_4$$

(3) By product HF and $CaSO_4$ can pose disposal problems.

Potassium fertilizer

Commonly potassium occurs as deposits of KCl.

(1) Mined as solid.
(2) Extracted through wells as KCl brine.

Fertilizer Pollution

(1) Runoff from crop fertilizer causes excessive algal growth in receiving waters resulting in eutrophicatio.

(2) Dead zone where Mississippi River enters Gulf of Mexico.

1) Reaches approximately 8000 square miles❶ area.
2) Lighter fresh water layer blankets saline water.
3) Laden with algal nutrients from agricultural fertilizers upstream.

磷肥

(1) P 以溶解性差的氟磷灰石矿物的形式出现,较少作为羟基磷灰石的形式出现。

1) $Ca_5F(PO_4)_3$;
2) $Ca_5OH(PO_4)_3$。

(2) 通过用磷酸或硫酸处理可将磷酸盐矿物转化为更易溶的过磷酸盐。

(3) 副产物 HF 和 $CaSO_4$ 可能造成处置问题。

钾肥

通常钾以 KCl 的沉积物形式存在。

(1) 以固体开采。
(2) 通过井提取为 KCl 盐水。

化肥污染

(1) 作物肥料的径流导致接受水域藻类过度生长,导致富营养化。

(2) 密西西比河进入墨西哥湾的死区。

1) 达到约 8000 平方英里❶的面积。
2) 淡水层覆盖了盐水。
3) 富含上游农业肥料中的藻类营养素。

---

❶ 1 square mile(平方英里)= 2589998.11m²。

4) Excessive phytoplankton growth produces biomass that decays.

① Consumes oxygen causing eutrophied water.

② Hydrogen sulfide produced.

4.2.9 Pesticides and their Residues in Soil 土壤中的农药及其残留

知识框架

(1) Four major categories of concern.

1) Carryover to later growing season.

2) Biological effects on organisms in terrestrial and aquatic ecosystems.

Bioaccumulation and transfer through food chains.

3) Groundwater contamination.

4) Effects on soil fertility.

(2) Herbicides are the most important soil contaminants.

Soil Fumigants

Soil fumigants are volatile substances applied to soil to combat harmful organisms largely on food crops.

Methyl bromide now banned.

4.2.10 Wastes and Pollutants in Soil 土壤中的废物和污染物

知识框架

(1) Soil receives wastes from the other

4) 浮游植物过度生长会导致生物量下降。

①消耗氧气导致富营养化的水。

②产生硫化氢。

(1) 值得关注的四个主要类别。

1) 延续到以后的生长季节。

2) 对陆地和水生生态系统中生物的影响。

通过食物链的生物富集和转移。

3) 地下水污染。

4) 对土壤肥力的影响。

(2) 除草剂是最重要的土壤污染物。

土壤熏蒸剂

土壤熏蒸剂是用于土壤的挥发性物质，主要用于对抗粮食作物上的有害生物。

现在禁止使用甲基溴。

(1) 土壤接收来自其他环

environmental spheres including those deliberately applied from the anthrosphere.

(2) Black carbon is an important repository of organic wastes.

May be left over from burning crop residues.

(3) Humic substances in soil tend to bind organic wastes.

(4) Pollutants degrade in soil.

1) Primarily biodegradation.

2) Rhizosphere where plant roots are located is a particularly active area for biodegradation.

3) Some chemical and photochemical degradation.

Soil Pollutants from Livestock Production

(1) Biodegradable biomass from wastes that consume oxygen in water and cause eutrophication.

(2) Fertilizer N, P, and O in wastes that cause excessive algal growth and eutrophication.

(3) In some cases, potentially harmful livestock feed additives.

1) Antibiotics.

2) Arsenic compounds such as Roxarsone were formerly used in poultry and swine feed.

境领域的废物，包括人类活动圈故意施加的物质。

（2）黑炭是有机废物的重要储存库。

可能残留有燃烧的农作物残渣。

（3）土壤中的腐殖质倾向于结合有机废物。

（4）污染物在土壤中降解。

1）主要是生物降解。

2）植物根系所在的根际是生物降解特别活跃的区域。

3）某些化学和光化学降解。

畜牧生产中的土壤污染物

（1）废物产生的可生物降解的生物质，消耗水中氧气并造成富营养化。

（2）废物中的肥料 N、P 和 O，导致藻类过度生长和富营养化。

（3）在某些情况下，可能有害的牲畜饲料添加剂。

1）抗生素。

2）砷化合物（如罗沙松）以前曾用于家禽和猪饲料。

重要名词和术语解析

Chemical degradation: Chemical degradation of pesticides has been observed experimentally in soils and clays sterilized to remove all microbial activity.

化学降解：在经过消毒以去除所有微生物活性的土壤和黏土中，已通过实验观察到农药的**化学降解**。

Photochemical reaction: A number of pesticides have been shown to undergo **photochemical reactions**, that is, chemical reactions brought about by the absorption of light.

Biodegradation: Although insects, earthworms, and plants may play roles in the **biodegradation** of pesticides and other pollutant organic chemicals, microorganisms have the most important role.

Rhizosphere: The **rhizosphere**, the layer of soil in which plant roots are especially active, is a particularly important part of soil with respect to biodegradation of wastes.

光化学反应：已证明许多农药发生**光化学反应**，即由光吸收引起的化学反应。

生物降解：尽管昆虫、蚯蚓和植物可能在农药和其他污染物有机化学物质的**生物降解**中起作用，但微生物起着最重要的作用。

根际：**根际**是植物根系特别活跃的土壤层，对于废物的生物降解而言，它是土壤中特别重要的部分。

4.2.11 Soil Loss and Degradation
水土流失和退化

知识框架

(1) Soil subject to degradation processes that make it unsuitable for crop growth.

(2) Soil erosion.

1) Water erosion has carried away millions of tons of topsoil.

2) Wind erosion.

Associated with formation of desert areas.

(3) Desertification.

1) Drought.

2) Fertility loss.

3) Vegetation loss.

4) Factors in desertification include:

①Erosion.

②Climate variations.

③Water availability.

（1）土壤经历退化过程，使其不适合作物生长。

（2）土壤侵蚀。

1）水蚀带走了数百万吨表土。

2）风蚀。

与沙漠地区的形成有关。

（3）荒漠化。

1）干旱。

2）生育力丧失。

3）植被丧失。

4）荒漠化的因素包括：

①侵蚀；

②气候变化；

③水的可用性；

④Loss of soil humus.
⑤Deterioration of chemical properties.
(4) Deforestation: Loss of forests.

1) Particular problem with tropical rain forests.
2) Major contributor to global warming.

Soil Sustainability and Water Resources
(1) Soil and water conservation are closely related.
(2) Precipitation falls on a land surface called a watershed.
(3) A properly maintained watershed.
1) Enables slow runoff.
2) Conserves water.
3) Prevents flooding.
4) Allows infiltration to aquifers.
(4) Permeable paving aids infiltration to groundwater in urban areas.

④土壤腐殖质流失；
⑤化学性质恶化。
（4）森林砍伐：森林的流失。

1) 热带雨林的特殊问题。

2) 导致全球变暖的主要因素。

土壤可持续性与水资源
（1）水土保持密切相关。
（2）降水落在称为分水岭的陆地上。
（3）妥善维护的分水岭。
1) 径流缓慢；
2) 保持水；
3) 防止洪水；
4) 允许渗入含水层。
（4）渗透性铺路有助于渗入城市地区的地下水。

重要名词和术语解析

Desertification: Desertification refers to the process associated with drought and loss of fertility by which soil becomes unable to grow significant amounts of plant life.
Deforestation: Loss of forests.
Soil conservation: The preservation of soil from erosion is commonly termed **soil conservation**.
Conservation tillage: For some crops **no-till agriculture**, now commonly called **conservation tillage**, greatly reduces erosion.

荒漠化：**荒漠化**是指与干旱和肥力丧失相关的过程，通过该过程土壤无法生长大量植物。

森林砍伐：森林的损失。
土壤养护：保护土壤免受侵蚀通常称为**土壤养护**。

保护性耕作：对于一些作物的**免耕农业**，现在称为**保护性耕作**，大大减少了侵蚀。

Watershed: The land area upon which rainwater falls is called a **watershed**.

分水岭：雨水流经的土地被称为**分水岭**。

4.2.12 Saving the Land
救治土地

知识框架

(1) Preventing soil erosion and deterioration is commonly called soil conservation.

(2) Soil is conserved by several important practices.

1) Terracing and contour plowing.

2) Planting cover crops.

3) Conservation tillage in which crops are planted in residue from the preceding season.

(3) Perennial plants conserve soil.

1) Grass for pasture and hay.

2) Trees.

3) A goal of genetic engineering in agriculture is development of perennial plants that produce cereal grains.

(1) 防止水土流失和退化通常被称为水土保持。

(2) 通过几种重要措施来保护土壤。

1) 梯田耕作和轮廓耕作。

2) 种植农作物。

3) 保护性耕作，将农作物种植上一季的残留物中。

(3) 多年生植物保护土壤。

1) 牧草和干草。

2) 树木。

3) 农业基因工程的目标是开发生产谷物的多年生植物。

Soil Restoration

(1) Soil can be impaired by loss of fertility, erosion, buildup of salinity, contamination by phytotoxins, such as zinc from sewage sludge, and other insults.

(2) Active measures of soil restoration, also called restoration ecology, may be required to restore productivity.

1) Physical alteration of soil to add terraces and relatively flat areas not subject erosion.

2) Restoration of organic matter by planting

土壤修复

(1) 肥力的丧失、侵蚀、盐分的积累、植物毒素（如污水污泥中的锌）的污染以及其他破坏可能会损害土壤。

(2) 要恢复生产力，可能需要采取积极的土壤恢复措施，也称为恢复生态学。

1) 物理改变土壤以增加不受侵蚀的梯田和相对平坦的区域；

2) 通过种植在土壤中培

high-biomass crops that are cultivated into the soil.

3) Nutrients added.

4) Contaminants treated.

①Excess acid or base neutralized.

②Salts leached from soil.

5) Removal of toxic substances by phytoremediation in which pollutants are pulled from soil by plant roots.

Poduculture in Soil Restoration

Poduculture is a means of growing crops in depleted soil, gravel, or sand that does not support plant growth well.

育的高生物量作物来恢复有机物质；

3) 添加营养；

4) 污染物处理。

①过量的酸或碱被中和。

②盐分从土壤中浸出。

5) 通过植物修复去除有毒物质，其中植物根系将污染物从土壤中拉出。

土壤修复中的荚式培育

荚式培育是一种在贫瘠的土壤、砾石或沙子等不能很好地支持植物生长的地方生长农作物的方法。

4.2.13 Green Chemistry and Sustainable Agriculture
绿色化学与可持续农业

知识框架

(1) Biomimetics applied to agriculture mimics natural systems to resist pests and increase productivity.

Largely applied to control pests.

(2) Biopesticides offer some advantages.

1) Lower toxicity.

2) High specificity.

3) Only small quantities required.

4) Rapid decomposition.

(3) Three classes of biopesticides.

1) Microbial pesticides such as insecticide-producing *Bacillus thuringiensis* bacteria.

(1) 用于农业的仿生植物模仿自然系统，以抵抗害虫并提高生产力。

广泛用于防治害虫。

(2) 生物农药具有一些优势。

1) 低毒性；

2) 高特异性；

3) 只需要少量；

4) 快速分解。

(3) 三类生物农药。

1) 微生物农药，例如产生杀虫剂的苏云金芽孢杆菌细菌。

2) Plant-incorporated protectants.

Plants genetically engineered to produce insecticide identical to that from *Bacillus thuringiensis* bacteria.

3) Nontoxic agents, such as sex attractants that confuse insects' reproductive activities

4.2.14 Genetics and Agriculture
遗传与农业

知识框架

(1) Ability to manipulate DNA in agricultural product DNA is causing huge changes in agricultural practice.

(2) The first Green Revolution of the mid-1960s used conventional selective breeding, hybridization, cross pollination, and back-crossing to vastly increase grain production and saved millions from starving.

1) New productive strains of rice, wheat, and corn.

2) Combined with use of chemical fertilizers.

(3) More recently, ability to manipulate DNA.

1) Crops resistant to weed-killing herbicides.

2) Crops that make their own insecticide.

(4) Future possibilities.

1) Much higher productivity.

2) Resistant to specific diseases.

3) Increased photosynthesis efficiency.

4) Capacity of non-leguminous plants to fix nitrogen.

2) 植物掺入的保护剂。

经过基因工程改造的植物产生的杀虫剂，与苏云金芽孢杆菌相同。

3) 会混淆昆虫生殖活动的无毒剂，例如引诱剂。

(1) 在农产品 DNA 中操纵 DNA 的能力正在引起农业实践的巨大变化。

(2) 1960 年代中期的第一次绿色革命，使用常规的选择性育种、杂交、异花授粉和回交来大幅增加谷物产量，并使数百万人免于饥饿。

1) 水稻、小麦和玉米的新生产菌株。

2) 结合使用化肥。

(3) 最近，操纵 DNA 的能力。

1) 抗除草剂的农作物。

2) 自己制造杀虫剂的农作物。

(4) 未来的可能性。

1) 更高的生产率。

2) 抗特定疾病。

3) 提高光合作用效率。

4) 非豆科植物固氮的能力。

4.2.15 Agriculture and Health
农业与健康

知识框架

（1）Health may have some relationship to land upon which crops and foods are produced.

1）Micronutrients, such as required low levels of selenium.

2）Possibility of toxic substances such as excessive selenium.

Geographic distribution of cancer may have some relationship to agriculture.

（2）Food contamination.

1）Contaminant pesticides may adversely affect health.

2）Contaminant bacteria (strains of *E. coli*) may cause disease.

3）Heavy metals, potentially from sewage sludge applied to land.

（1）健康可能与生产农作物和粮食的土地有关。

1）微量营养素，如所需的低含量硒。

2）例如过量硒成为有毒物质的可能性。

癌症的地理分布可能与农业有关。

（2）食物污染。

1）污染物农药可能对健康产生不利影响。

2）污染细菌（大肠杆菌菌株）可能导致疾病。

3）重金属，可能来自应用于土地的污水污泥。

4.2.16 Protecting the Food Supply from Attack
保护粮食供应不受攻击

知识框架

（1）Large-scale chemical attack on crops would be almost impossible.

（2）Pathogens could be introduced into food.

Each year people are sickened by inadvertent contamination of food by pathogens such as *Salmonella* or disease-causing strains of *E. coli*.

（3）Crop and livestock pathogen attacks could occur.

Major example is livestock anthrax that could affect humans.

（1）对作物进行大规模的化学攻击几乎是不可能的。

（2）病原体可以引入食物。

每年，人们都会因诸如沙门氏菌或致病性大肠杆菌等病原体对食品的无意污染而感到恶心。

（3）病原体攻击作物和牲畜可能发生。

例：可能影响人类的牲畜炭疽

Exercises
习题

4-1 Give two examples of reactions involving manganese and iron compounds that may occur in waterlogged soil.

4-2 What temperature and moisture conditions favor the buildup of organic matter in soil?

4-3 "Cat clays" are soils containing a high level of iron pyrite, FeS_2. Hydrogen peroxide, H_2O_2, is added to such a soil, producing sulfate as a test for cat clays. Suggest the chemical reaction involved in this test.

4-4 What effect upon soil acidity would result from heavy fertilization with ammonium nitrate accompanied by exposure of the soil to air and the action of oxic bacteria?

4-5 How many moles of H^+ ion are consumed when 200kg of $NaNO_3$ undergo denitrification in soil?

4-6 What is the primary mechanism by which organic material in soil exchanges cations?

4-7 Prolonged waterlogging of soil does not (　　).

　A. increase NO_3^- production

　B. increase Mn^{2+} concentration

　C. increase Fe^{2+} concentration

　D. have harmful effects upon most plants

　E. increase production of NH_4^+ from NO_3^-

4-8 Of the following phenomena, the one that eventually makes soil more basic is (　　).

4-1 举两个例子，说明在积水的土壤中可能发生的涉及锰和铁化合物的反应。

4-2 什么温度和湿度条件有利于土壤中有机物的堆积？

4-3 "猫砂"是指含有大量黄铁矿 FeS_2 的土壤。在这样的土壤中加入过氧化氢 H_2O_2，生成硫酸盐，作为猫砂的试验。写出该试验中的化学反应。

4-4 如果大量施用硝酸铵，同时将土壤暴露在空气中，再加上氧化细菌的作用，会对土壤酸度产生什么影响？

4-5 200kg $NaNO_3$ 在土壤中进行反硝化作用，要消耗多少摩尔的 H^+ 离子？

4-6 土壤中有机质交换阳离子的主要机制是什么？

4-7 土壤长期积水不会（　　）。

　A. 增加 NO_3^- 的产生

　B. 增加 Mn^{2+} 浓度

　C. 增加 Fe^{2+} 浓度

　D. 对大多数植物有有害影响

　E. 增加 NO_3^- 产生的 NH_4^+

4-8 下列现象中，最终使土壤碱性增强的是（　　）。

A. removal of metal cations by roots
B. leaching of soil with CO_2-saturated water
C. oxidation of soil pyrite
D. fertilization with $(NH_4)_2SO_4$
E. fertilization with KNO_3

4-9 How many metric tons of farm manure are equivalent to 100kg of 10-5-10 fertilizer?

4-10 How are the chelating agents that are produced from soil microorganisms involved in soil formation?

4-11 What specific compound is both an animal waste product and a major synthetic fertilizer?

4-12 What happens to the nitrogen/carbon ratio as organic matter degrades in soil?

4-13 To prepare a rich potting soil, a greenhouse operator mixed 75% "normal" soil with 25% peat. Estimate the cation-exchange capacity in 10^{-2} mmol/g of the product.

4-14 Explain why plants grown on either excessively acidic or excessively basic soils may suffer from calcium deficiency.

4-15 What are two mechanisms by which anions may be held by soil mineral matter?

4-16 What are the three major ways in which pesticides are degraded in or on soil?

4-17 Lime from lead mine tailings containing 0.5% (by mass) lead was applied at a rate of 10t per acre of soil and worked in to a depth of 20cm. The soil density was

A. 根系对金属阳离子的去除
B. CO_2 饱和水对土壤的浸染
C. 土壤黄铁矿的氧化
D. 用 $(NH_4)_2SO_4$ 施肥
E. 用 KNO_3 施肥

4-9 多少吨农家肥相当于100kg 10-5-10肥料?

4-10 土壤微生物产生的螯合剂是如何参与土壤形成的?

4-11 什么特定的化合物既是动物废弃物,又是主要的合成肥料?

4-12 随着土壤中有机物的降解,氮/碳比会发生什么变化?

4-13 为了制备一种肥沃的盆栽土壤,某温室经营者将75%的"普通"土壤与25%的泥炭混合。估计阳离子交换能力,单位为 10^{-2} mmol/g 产品。

4-14 解释为什么在酸性过强或碱性过强的土壤上生长的植物会缺钙。

4-15 阴离子可能被土壤矿物质保持的两种机制是什么?

4-16 农药在土壤中或土壤上降解的三大途径是什么?

4-17 从含铅量(质量分数)为0.5%的铅矿尾矿中按每英亩❶ 10t的速度施用石灰,施用深度为20cm。

---

❶ 1acre(英亩)= 4046.856m^2。

2.0g/cm³. To what extent did this add to the burden of lead in the soil? (There are 640 acres per square mile and 1,609m per mile.)

4-18 Match the soil or soil-solution constituent in the number column with the soil condition described on the letter column, below.

(1) High Mn^{2+} content in soil
(2) Excess H^+
(3) High H^+ and SO_4^{2-} concentrations
(4) High organic content

(A) "Cat clays" containing initially high levels of pyrite, FeS_2
(B) Soil in which biodegradation has not occurred to a great extent
(C) Waterlogged soil
(D) Soil, the fertility of which can be improved by adding limestone

4-19 What are the processes occurring in soil that operate to reduce the harmful effects of pollutants?

4-20 Under what conditions do the reactions below occur in soil? Name two detrimental effects that can result from these reactions.

(1) $MnO_2 + 4H^+ + 2e^- \rightarrow Mn^{2+} + 2H_2O$.
(2) $Fe_2O_3 + 6H^+ + 2e^- \rightarrow 2Fe^{2+} + 3H_2O$.

土壤密度为 2.0g/cm³。这在多大程度上加重了土壤中铅的负担？（每平方英里有640英亩，每英里有1609m）

4-18 请将数字栏中的土壤或土壤溶液成分，与字母栏中的土壤条件相匹配。

(1) 土壤中 Mn^{2+} 含量高
(2) 过量的 H^+
(3) 高 H^+ 和 SO_4^{2-} 浓度
(4) 有机物含量高

(A) 最初含有高含量黄铁矿 FeS_2 的"猫砂"
(B) 尚未发生严重生物降解的土壤
(C) 浸水的土壤
(D) 土壤，其肥力可以通过添加石灰石来改善

4-19 土壤中发生的哪些过程可以减少污染物的有害影响？

4-20 下列反应在土壤中是在什么条件下发生的？请说出这些反应可能产生的两种不利影响。

(1) $MnO_2 + 4H^+ + 2e^- \rightarrow Mn^{2+} + 2H_2O$；
(2) $Fe_2O_3 + 6H^+ + 2e^- \rightarrow 2Fe^{2+} + 3H_2O$。

4-21 What are four important effects of organic matter in soil?

4-22 How might irrigation water treated with fertilizer containing potassium and ammonia become depleted of these nutrients in passing through humus-rich soil?

4-21 土壤中有机质的四个重要作用是什么？

4-22 用含钾和氨的化肥处理过的灌溉水在通过富含腐殖质的土壤时，这些营养物质会如何被消耗掉？

扫描二维码查看
习题答案

5 Environmental Chemistry of the Biosphere
生物圈环境化学

5.1 The Biosphere: Environmental Biochemistry
生物圈：环境生物化学

5.1.1 Life and the Biosphere
生命和生物圈

知识框架

(1) All living organisms.

1) Process matter and energy through the process of metabolism.

2) Are capable of reproduction.

(2) Two other important characteristics are:

1) Maintenance of internal temperature and other conditions. Mammals and other warm-blooded animals maintain constant body temperature.

2) Undergo changes in genetic characteristics through generations that lead to better survivability.

(3) Life requires several things from the Earth System.

1) Liquid water.

2) Essential chemical elements.

3) Temperatures within a suitable relatively stable range.

4) Sheltering atmosphere.

①O_2 for animals.

(1) 所有的生物。

1) 通过代谢过程处理物质和能量。

2) 有繁殖能力。

(2) 两个其他重要特征是：

1) 维持内部温度和其他条件，哺乳动物和其他温血动物保持恒定的体温；

2) 世代相传的遗传特性变化导致更好的生存能力。

(3) 生命需要地球系统的几样东西。

1) 液态水。

2) 基本化学元素。

3) 温度在合适的相对稳定范围内。

4) 庇护的大气。

①供给动物的 O_2；

② CO_2 for plants.

3) Relatively free of toxic substances.

Nature of Life Determined by Surroundings

(1) Abiotic external factors.

(2) Aquatic or terrestrial environment.

(3) For the terrestrial environment.

1) Suitable soil.

2) Available water.

3) Essential nutrients.

(4) Biotic external factors.

1) Other life forms present.

2) Predatory, non-predatory, or parasitic.

3) Source of food.

4) Wastes.

The Biosphere in Stabilizing the Earth System: The Gaia Hypothesis

Organisms largely determine and maintain Earth's climate.

Atmospheric O_2/CO_2 balance through photosynthesis.

② 供给植物的 CO_2。

3）相对不含有毒物质。

生命的本质取决于周围的环境

（1）非生物外部因素。

（2）水生或陆地环境。

（3）对于陆地环境。

1）合适的土壤。

2）足够的水。

3）关键营养物。

（4）生物外部因素。

1）存在其他生命形式。

2）掠食性，非掠食性或寄生性。

3）食物的来源。

4）废弃物。

稳定地球系统中的生物圈：盖亚假说

生物在很大程度上决定和维持地球的气候。

通过光合作用维持大气 O_2/CO_2 平衡。

重要名词和术语解析

Biosphere: The **biosphere** is where living organisms function in the water-rich boundary region at the interface of Earth's surface with the atmosphere, a very thin region compared to the dimensions of Earth or its atmosphere.

Gaia hypothesis: **Gaia hypothesis** contends that the atmospheric O_2/CO_2 balance established and sustained by organisms determines

生物圈：生物圈是生物体在地球表面与大气之间的界面处的富水边界区域中起作用的区域，与地球或其大气层的尺寸相比，这是一个非常薄的区域。

盖亚假说：盖亚假说认为，生物建立并维持的大气 O_2/CO_2 平衡，主要通过光合作用决定

and maintains Earth's climate and other environmental conditions, largely through photosynthesis.

5.1.2 Metabolism and Control in Organisms
生物体内的代谢与控制

并维持地球的气候和其他环境条件。

知识框架

(1) Metabolism is the mechanism by which organisms process matter and energy.

1) Photosynthesis to produce biomass, represented as $\{CH_2O\}$, that forms the base of the food chain:

$$CO_2 + H_2O + h\nu \longrightarrow \{CH_2O\} + O_2$$

2) Digestion to break down complex food molecules.

3) Respiration for energy from food (glucose):

$$C_6H_{12}O_6 \text{ (glucose, 葡萄糖)} + 6O_2 \longrightarrow 6CO_2 + 6H_2O + energy, 能量$$

(2) Organisms assemble small molecules to produce biomolecules, such as proteins, by a synthesis process.

(3) Enzymes are biochemical catalysts that carry out metabolism.

1) Make reactions much faster.

2) Make reactions specific.

3) Affected by factors such as temperature.

(4) Nutrients required for metabolism.

(1) 代谢是生物体处理物质和能量的机制。

1) 光合作用产生生物量，表示为 $\{CH_2O\}$，形成食物链的基础：

2) 消化分解复杂的食物分子。

3) 呼吸从食物（葡萄糖）获得能量：

(2) 有机体通过合成过程组装小分子以产生生物分子，例如蛋白质。

(3) 酶是进行新陈代谢的生化催化剂。

1) 使反应更快。

2) 使反应更明确。

3) 受温度等因素的影响。

(4) 生物体代谢需要营养素。

Control and Regulation in Organisms

(1) Maintenance of homeostasis.

(2) Control through nervous system.

(3) Nerve impulses (rapid).

Control through messenger molecules (hormones).

生物体的控制与调节

(1) 维持体内平衡。

(2) 通过神经系统控制。

(3) 神经冲动（快速）。

通过信使分子（激素）进行控制。

重要名词和术语解析

Metabolism: The ability of an organism to process matter and energy is called **metabolism**.

Photosynthesis: Photosynthesis is the metabolic process that provides the base of the food chain for most organisms.

Digestion: Animals break down complex food materials to smaller molecules through the process of digestion.

Respiration: Respiration occurs as nutrients are metabolized to yield energy.

Synthesis: Organisms assemble small molecules to produce biomolecules, such as proteins, by a synthesis process.

Nutrient: The raw materials that organisms require for their metabolism are nutrients.

Nerve impulse: The most obvious means of control in animals is through the nervous system in which messages are conducted very rapidly to various parts of the animal as nerve impulses.

Central nervous system: More advanced animals have a brain and spinal cord that function as a central nervous system.

Peripheral nenropathy: Exposure to organic solvents that dissolve some of the protective lipids around nerve fibers can lead to a condition in which limbs do not function properly, called peripheral nenropathy.

新陈代谢：生物体处理物质和能量的能力称为**新陈代谢**。

光合作用：光合作用是为大多数生物提供食物链基础的代谢过程。

消化：动物通过消化过程将复杂的食物分解成较小的分子。

呼吸：呼吸发生在营养物质代谢产能时。

合成：有机体通过合成过程组装小分子以产生生物分子，如蛋白质。

营养素：生物体代谢所需的原材料是营养素。

神经冲动：动物最明显的控制手段是通过神经系统，在该系统中，作为神经冲动的信息很快传递到动物的各个部位。

中枢神经系统：较高级的动物具有充当中枢神经系统的大脑和脊髓。

周围性神经病：暴露于能溶解神经纤维周围一些保护性脂质的有机溶剂中会导致肢体功能不正常的情况，称为周围性神经病。

5.1.3 Reproduction and Inherited Traits
繁殖与遗传性状

知识框架

(1) Reproduction is an essential function of organisms.

1) Over-reproduction causes problems.

2) Potential effects of environmental pollutants to stop reproduction and produce defective progeny.

(2) Reproduction is directed by genes which occur in molecules of deoxyribonucleic acid, DNA (see discussion of biochemistry).

(3) Mutations that may be caused by mutagens in food and the environment.

(1) 繁殖是生物的基本功能。

1) 过度繁殖会引起问题。

2) 环境污染物可能阻止繁殖,以及产生有缺陷的后代。

(2) 繁殖由脱氧核糖核酸DNA分子中存在的基因指导。

(3) 食物和环境中的诱变剂可能引起突变。

5.1.4 Stability and Equilibrium of Biosphere
生物圈的稳定和平衡

知识框架

(1) Homeostasis in which an organism reaches a state of stability and equilibrium with its environment.

1) Temperature.

2) Internal levels of water, salts, calcium.

(2) Homeostasis applies to individuals and populations.

(3) Ecology describes the interaction of organisms with their surroundings and each other.

(4) An ecosystem is a segment of the environment and organisms in it with their various interactions and relationships.

(5) Factors in ecosystems include light,

(1) 生物体与其环境达到稳定和平衡状态的内稳态。

1) 温度。

2) 内部水、盐、钙的含量。

(2) 内稳态适用于个体和种群。

(3) 生态学描述了生物与其周围环境以及彼此之间的相互作用。

(4) 生态系统是环境和生物在其中的各种相互作用和关系的一部分。

(5) 生态系统中的因素包

temperature, water, other organisms.

(6) Food chains and food webs are important aspects of ecosystems.

(7) Biomagnification of environmental pollutants in ecosystems.

括光、温度、水、其他生物。

(6) 食物链和食物网是生态系统的重要方面。

(7) 生态系统中环境污染物的生物放大。

重要名词和术语解析

<u>Homeostasis</u>: In order for an organism to survive and thrive, it must reach a state of stability and equilibrium with its environment. The term given to such a state is homeostasis ("same status").

<u>Ecology</u>: Ecology describes the interaction of organisms with their surroundings and each other.

<u>Ecosystem</u>: An ecosystem describes a segment of the environment and the organisms in it with all of the interactions and relationships that implies.

<u>Food chain</u> or <u>food web</u>: An important part of any ecosystem is the food chain, or more complicated food webs, in which food generated by photosynthesis is utilized by different organisms at different levels.

<u>Biomagnification</u>: Aquatic pollutants become more concentrated in lipid tissue at the top of the food chain, a process called biomagnification.

内稳态：为了使生物能够生存和繁衍，必须使其与环境保持稳定和平衡的状态。赋予这种状态的术语是内稳态（"相同状态"）。

生态学：生态学描述了生物与其周围环境以及彼此之间的相互作用。

生态系统：生态系统描述了环境和其中的有机体的一部分，并暗示了所有相互作用和关系。

食物链或**食物网**：任何生态系统的重要组成部分是食物链，或更复杂的食物网，其中光合作用产生的食物在不同水平被不同生物利用。

生物放大：水生污染物在食物链顶部的脂质组织中的浓度更高，这一过程称为生物放大。

5.1.5 Biochemistry
生物化学

知识框架

(1) Biochemistry is the chemistry of living

(1) 生物化学是生物系统

systems.

1) Chemical properties.

2) Chemical composition.

3) Processes of complex substances in living systems.

(2) Environmental biochemistry deals with biochemical phenomena in the environment.

1) Soil.

1) Water.

(3) Biomolecules.

1) Macromolecules such as proteins.

2) Small molecules such as glucose sugar.

3) Carbohydrates.

4) Proteins.

5) Lipids.

6) Nucleic acids.

7) Hydrophilic molecules such as glucose sugar.

8) Hydrophobic molecules such as lipid (fat) molecules.

的化学。

1) 化学性质。

2) 化学成分。

3) 生命系统中复杂物质的过程。

(2) 环境生物化学处理环境中的生物化学现象。

1) 土壤。

2) 水。

(3) 生物分子。

1) 大分子，例如蛋白质。

2) 小分子，例如葡萄糖。

3) 碳水化合物。

4) 蛋白质。

5) 脂质。

6) 核酸。

7) 亲水性分子，例如葡萄糖。

8) 疏水性分子，例如脂质（脂肪）分子。

重要名词和术语解析

Biochemistry：**Biochemistry** is the chemistry that deals with the chemical properties, composition, and biologically-mediated processes of complex substances in living systems.

Environmental biochemistry：Biochemical processes not only are profoundly influenced by chemical species in the environment, they largely determine the nature of these species, their degradation, and even their syntheses, particularly in the aquatic and soil environments. The study of such phenomena forms the basis of **environmental biochemistry**.

生物化学：生物化学是处理生命系统中复杂物质的化学性质、组成和生物介导过程的化学。

环境生物化学：生化过程不仅受到环境中化学物种的深刻影响，而且在很大程度上决定了这些物种的性质，其降解甚至合成，特别是在水生和土壤环境中。对这种现象的研究构成了**环境生物化学**的基础。

5.1.6 Biochemistry and the Cell
生物化学和细胞

知识框架

(1) Prokaryotic cells do not have a defined nucleus.

Primarily in single-celled organisms, especially bacteria.

(2) Eukaryotic cells have a nucleus.

In higher, often multicelled life forms including plants, animals, and fungi.

Major Cell Features(of eukaryotic cell)

(1) Cell membrane.
1) Encloses cell.
2) Regulates passage of materials.
(2) Nucleus.
1) Cell control.
2) Contains DNA.
3) Reproduction.
(3) Cytoplasm.
Fills cell interior except for nucleus, organelles, vacuoles.
(4) Mitochondria.
Mediate energy conversion and utilization.
(5) Ribosomes: Involved in protein synthesis.

(6) Endoplasmic reticulum: Sites of metabolism of some toxicants.
(7) Lysosome: Involved in food digestion.

(8) Golgi bodies: Contain substances produced by cells.

(1) 原核细胞没有明确的细胞核。

主要存在于单细胞生物中，尤其是细菌。

(2) 真核细胞有细胞核。

以较高的，通常是多细胞的生命形式存在，包括植物、动物和真菌。

真核细胞的主要细胞特征

(1) 细胞膜。
1) 封闭细胞。
2) 调节物料的通过。
(2) 细胞核。
1) 细胞控制。
2) 包含 DNA。
3) 繁殖。
(3) 细胞质。
除细胞核、细胞器、液泡外，填充细胞内部。
(4) 线粒体。
中介能源转换和利用。
(5) 核糖体：参与蛋白质合成。

(6) 内质网：某些有毒物质的代谢位点。
(7) 溶酶体：参与食物消化。

(8) 高尔基体：包含细胞产生的物质。

(9) Cell walls: Stiffness and structure.

(10) Vacuoles: Often contain materials dissolved in water.

(11) Chloroplasts: Sites of photosynthesis in plant cells.

(9) 细胞壁：刚度和结构。

(10) 液泡：通常包含溶于水的物质。

(11) 叶绿体：植物细胞中光合作用的位点。

重要名词和术语解析

Cell: The basic building block of living systems where most life processes are carried out.

Prokaryotic: Prokaryotic cells do not have a defined nucleus.

Eukaryotic: Eukaryotic cells have a nucleus.

Cell membrane: Cell membrane encloses the cell and regulates the passage of ions, nutrients, lipid-soluble ("fat-soluble") substances, metabolic products, toxicants, and toxicant metabolites into and out of the cell interior because of its varying **permeability** for different substances.

Cell nucleus: Cell nucleus acts as a sort of "control center" of the cell. It contains the genetic directions the cell needs to reproduce itself. The key substance in the nucleus is **deoxyribonucleic acid** (**DNA**).

Chromosomes: Chromosomes in the cell nucleus are made up of combinations of DNA and proteins.

Cytoplasm: Cytoplasm fills the interior of the cell not occupied by the nucleus. Cytoplasm is further divided into a water-soluble proteinaceous filler called **cytosol**, in which are suspended bodies called **cellular organelles**, such as

细胞：是进行大多数生命过程的生命系统的基本组成部分。

原核：原核细胞没有明确的细胞核。

真核生物：真核细胞有细胞核。

细胞膜：细胞膜包围细胞，并由于不同的物质具有不同的**渗透性**，调节离子、营养物质、脂溶性物质、代谢产物、有毒物质和有毒代谢产物的进入和流出细胞内部。

细胞核：细胞核是细胞的"控制中心"。它包含细胞繁殖自身所需的遗传指导。细胞核中的关键物质是**脱氧核糖核酸**（**DNA**）。

染色体：细胞核中的染色体由 DNA 和蛋白质的组合组成。

细胞质：细胞质填充未被细胞核占据的细胞内部。细胞质进一步分为称为细胞质的水溶性蛋白质填充物，其中是称为**细胞器的悬浮体**，例如线粒体，

mitochondria or, in photosynthetic organisms, chloroplasts.

Mitochondria: Mitochondria are the "powerhouses" which mediate energy conversion and utilization in the cell.

Ribosome: Ribosomes participate in protein synthesis.

Endoplasmic reticulum: Endoplasmic reticulum is involved in the metabolism of some toxicants by enzymatic processes.

Lysosome: Lysosome is a type of organelle that contains potent substances capable of digesting liquid food material. Such material enters the cell through a "dent" in the cell wall, which eventually becomes surrounded by cell material. This surrounded material is called a **food vacuole**. The digestion process consists largely of **hydrolysis reactions** in which large, complicated food molecules are broken down into smaller units by the addition of water.

Golgi bodies: Golgi bodies occur in some types of cells. These are flattened bodies of material that serve to hold and release substances produced by the cells.

Cell walls: Cell walls of plant cells are strong structures that provide stiffness and strength.

Vacuoles: Vacuoles inside plant cells that often contain materials dissolved in water.

Chloroplasts: Chloroplasts in plant cells that are involved in photosynthesis (the chemical process which uses energy from sunlight to convert carbon dioxide and water to organic matter). Photosynthesis occurs in these

在光合生物中为叶绿体。

线粒体：线粒体是调节细胞能量转化和利用的"动力源"。

核糖体：核糖体参与蛋白质合成。

内质网：内质网通过酶促过程参与某些有毒物质的代谢。

溶酶体：溶酶体是一种细胞器，其中含有能够消化液态食品原料的有效物质。这种材料通过细胞壁上的"凹痕"进入细胞，最终被细胞材料包围。这种被包围的材料称为**食物泡**。消化过程主要由**水解反应**组成，通过加成水将大而复杂的食物分子分解成较小的单元。

高尔基体：高尔基体存在于某些类型的细胞中。这些扁平的材料体用于保持和释放细胞产生的物质。

细胞壁：植物细胞的细胞壁是坚固的结构，可提供刚度和强度。

液泡：植物细胞内部的液泡，通常含有溶解在水中的物质。

叶绿体：植物细胞中的叶绿体参与光合作用（一种化学过程，利用阳光中的能量将二氧化碳和水转化为有机物）。这些体内发生光合作用。通过光

bodies. Food produced by photosynthesis is stored in the chloroplasts in the form of **starch grains**.

5.1.7 Proteins
蛋白质

知识框架

(1) Proteins are nitrogen-containing macromolecules that are basic units of live systems.

1) Cytoplasm largely protein.

2) Enzymes (biological catalysts) are proteins.

(2) The protein macromolecule is composed of hundreds or thousands of amino acid molecules.

1) Each amino acid joined to protein with loss of H_2O through the peptide linkage.

2) General formula of amino acids where R is an organic group or, in the case of glycine, H.

Major Types of Proteins

(1) Nutrient (casein), food sources.

(2) Storage (ferritin), iron storage.

(3) Structural (collagen), tendons and hair.

(4) Contractile (myosin), muscle movement.

(5) Transport (hemoglobin), move species such as O_2.

(6) Defense, antibodies against foreign agents.

合作用产生的食物以淀粉颗粒的形式存储在**叶绿体**中。

(1) 蛋白质是含氮大分子，是生命系统的基本单位。

1) 细胞质主要是蛋白质。

2) 酶（生物催化剂）是蛋白质。

(2) 蛋白质大分子由数百或数千个氨基酸分子组成。

1) 每个氨基酸通过肽键与蛋白质结合而失去 H_2O。

2) 氨基酸的通式，其中 R 是有机基团；对于甘氨酸，则为 H。

蛋白质的主要类型

(1) 营养（酪蛋白），食物来源。

(2) 储存（铁蛋白），铁储存。

(3) 结构（胶原蛋白），肌腱和头发。

(4) 收缩（肌球蛋白），肌肉运动。

(5) 运输（血红蛋白），移动如 O_2 之类的物质。

(6) 防御，针对外来因子的抗体。

(7) Regulatory (insulin), regulate biochemical processes.

(8) Enzymes (acetylcholinesterase), biochemical reaction catalysts.

Protein Structure

(1) Primary: Order of amino acids.

(2) Secondary: Bonding between polypeptide protein chains.

(3) Tertiary: Twisting of alpha helix secondary structures into specific shapes.

Determine shape of protein molecule and how it is recognized by other molecules such as proteinaceous enzymes.

(4) Quaternary: Formed by two or more protein molecules attracted together.

Fibrous proteins (hair, muscle, tendons).

Water-insoluble.

Globular proteins (enzymes, hemoglobin).

Denaturation of Proteins (destruction of structure)

Secondary, tertiary, and quaternary protein structures are easily changed by a process called denaturation.

(1) Heating.

(2) Violent agitation.

(3) Toxic agents.

Denaturation is a major mode of the action of toxicants.

Example: Lead binding to S in protein.

（7）监管（胰岛素），监管生化过程。

（8）酶（乙酰胆碱酯酶），生化反应催化剂。

蛋白质结构

（1）一级：氨基酸顺序。

（2）二级：多肽蛋白链之间的键合。

（3）三级：将 α 螺旋二级结构扭曲成特定形状。

确定蛋白质分子的形状以及如何被蛋白质酶等其他分子识别。

（4）四级：由两个或多个蛋白质分子吸引在一起形成。

纤维蛋白（头发，肌肉，肌腱）。

不溶于水。

球状蛋白（酶，血红蛋白）。

蛋白质变性（破坏结构）

二级，三级和四级蛋白质的结构很容易通过变性过程改变。

（1）加热。

（2）剧烈搅拌。

（3）有毒物质。

变性是毒物作用的主要方式。

例：铅与蛋白质中的 S 结合。

重要名词和术语解析

Proteins: Proteins are nitrogen-containing organic compounds which are the basic units of live systems.

Amino acids: Proteins are made up of **amino acids** joined together in huge chains. Amino acids are organic compounds which contain the carboxylic acid group, $-CO_2H$, and the amino group, $-NH_2$.

Peptide linkages: Proteins are polymers or **macromolecules** of amino acids containing from approximately 40 to several thousand amino acid groups joined by **peptide linkages**.

Polypeptides: Smaller molecule amino acid polymers, containing only about 10 to about 40 amino acids per molecule, are called **polypeptides**.

Residue: A portion of the amino acid left after the elimination of H_2O during polymerization is called a **residue**.

Protein structure: The order of amino acids in protein molecules, and the resulting three-dimensional structures that form, provide an enormous variety of possibilities for protein structure.

Primary structure: The order of amino acids in the protein molecule determines its primary structure.

Secondary structure: Secondary protein structures result from the folding of polypeptide protein chains to produce a maximum number of hydrogen bonds between peptide linkages.

Alpha-helix: With larger R groups the molecules tend to take a spiral form. Such a

蛋白质：蛋白质是含氮有机化合物，是生命系统的基本单位。

氨基酸：蛋白质是由**氨基酸**组成的长链。氨基酸是含有羧基—CO_2H 和氨基—NH_2 的有机化合物。

肽键：蛋白质是氨基酸的高分子或**大分子**，包含约 40 至数千个通过肽键连接的**氨基酸基团**。

多肽：每个分子仅包含约 10~40 个氨基酸的小分子氨基酸聚合物称为**多肽**。

残基：聚合过程中消除 H_2O 后剩下的一部分氨基酸称为**残基**。

蛋白质结构：蛋白质分子中氨基酸的顺序以及形成的三维结构为蛋白质结构提供了多种可能性。

一级结构：蛋白质分子中氨基酸的顺序决定了其一级结构。

二级结构：二级蛋白质结构是多肽蛋白质链折叠产生的，在肽键之间产生最大数量的氢键。

α-螺旋：具有较大的 R 基团的分子倾向于呈螺旋形式。这样

spiral is known as an alpha-helix.
Tertiary structure: Tertiary structures are formed by the twisting of alpha-helices into specific shapes.
Quaternary structure: Two or more protein molecules consisting of separate polypeptide chains may be further attracted to each other to produce a quaternary structure.
Fibrous proteins: Fibrous proteins occur in skin, hair, wool, feathers, silk, and tendons.

Globular protein: Globular protein is the other major type of protein form aside from fibrous protein.
Denaturation: Secondary, tertiary, and quaternary protein structures are easily changed by a process called **denaturation**.

5.1.8 Carbohydrates
碳水化合物

的螺旋被称为 α-螺旋。
三级结构：三级结构是通过将 α-螺旋扭曲成特定形状而形成的。
四级结构：两个或更多由单独的多肽链组成的蛋白质分子可以进一步相互吸引，以产生四级结构。
纤维蛋白：纤维蛋白存在于皮肤、头发、羊毛、羽毛、丝绸和肌腱中。
球蛋白：除纤维蛋白外，球蛋白是另一种主要的蛋白质形式。
变性：通过称为**变性**的过程很容易改变二级、三级和四级蛋白质的结构。

知识框架

(1) Approximate simple formula CH_2O.

(2) Small-molecule carbohydrates such as glucose.

(3) Large-molecule carbohydrates such as starch.
1) Macromolecule composed of several hundred glucose units.
2) A food source, can be digested by humans.
(4) Carbohydrates with very large molecules such as cellulose.
1) Not digested directly by humans.

（1）近似简单化学式 CH_2O。

（2）葡萄糖等小分子碳水化合物。

（3）淀粉等大分子碳水化合物。
1）由数百个葡萄糖单位组成的大分子。
2）食物来源，人类可以消化。
（4）具有非常大的分子的碳水化合物，如纤维素。
1）不被人类直接消化。

2) Digested by bacteria such as those in stomachs of ruminant animals.

2) 被反刍动物的胃中的细菌消化。

重要名词和术语解析

Carbohydrates: Carbohydrates have the approximate simple formula CH_2O and include a diverse range of substances composed of simple sugars such as glucose.

Polysaccharides: High-molar-mass polysaccharides, such as starch and glycogen ("animal starch"), are biopolymers of simple sugars.

Monosaccharide: The simplest carbohydrates are the **monosaccharides**, also called simple sugars. Units of two monosaccharides make up several very important sugars known as **disaccharides**.

Polysaccharide: Polysaccharides consist of many simple sugar units hooked together. One of the most important polysaccharides is **starch**, which is produced by plants for food storage. Animals produce a related material called **glycogen**.

Cellulose: Cellulose is a polysaccharide which is also made up of $C_6H_{10}O_5$ units.

Glycoprotein: Carbohydrate groups are attached to protein molecules in a special class of materials called **glycoproteins**.

碳水化合物：碳水化合物的化学式近似为 CH_2O，并包含由单糖（如葡萄糖）组成的多种物质。

多糖：淀粉和糖原（"动物淀粉"）等高摩尔质量，多糖是单糖的生物聚合物。

单糖：最简单的碳水化合物是单糖。两个单糖的单位组成了几个非常重要的糖，称为**二糖**。

多糖：多糖由许多相互连接的简单糖单元组成。最重要的多糖之一是**淀粉**，淀粉是由植物生产用于食品储存的。动物产生一种称为**糖原**的相关物质。

纤维素：纤维素是一种多糖，也由 $C_6H_{10}O_5$ 单元组成。

糖蛋白：碳水化合物基团在称为**糖蛋白**的特殊材料中与蛋白质分子连接。

5.1.9 Lipids
脂质

知识框架

(1) Lipids are biomolecules extractable from tissue into organic solvents such as diethyl ether or dichloromethane.

(1) 脂质是可从组织中提取到有机溶剂（如乙醚或二氯甲烷）中的生物分子。

(2) Lipids include a large variety of biomaterials.

1) Triglyceride esters are the most common.

①Fats.
②Vegetable oils.
2) Waxes.
3) Cholesterol.
4) Some vitamins.
5) Some hormones.
6) Organic-soluble long-chain fatty acids.

(3) Lipids serve a variety of functions:
1) Food storage.
2) Protective coatings (especially waxes).
3) Steroids, some of which are hormones.
4) Phosphoglycerides in cell membranes.

(4) Lipids are important in toxicology:

1) Some toxic substances interfere with lipid metabolism, which can cause lipid accumulation ("fatty liver").

2) Organophilic toxicants such as PCBs become concentrated in lipid tissue.

(5) Waxes are esters of alcohols and fatty acids.

(6) Some steroids are hormones, which act as messengers in organisms.

（2）脂质包括多种生物材料。

1）甘油三酸酯是最常见的。

①脂肪；
②植物油。
2）蜡。
3）胆固醇。
4）一些维生素。
5）一些激素。
6）有机可溶性长链脂肪酸。

（3）脂质具有多种功能：
1）食品储存。
2）保护性涂层（尤其是蜡）。
3）类固醇，其中一些是激素。
4）细胞膜中的磷酸甘油酯。

（4）脂质在毒理学中很重要。

1）一些有毒物质会干扰脂质代谢，从而导致脂质堆积（"脂肪肝"）。

2）多氯联苯等亲有机毒物集中在脂质组织中。

（5）蜡是醇和脂肪酸的酯。

（6）一些类固醇是激素，在有机体中充当信使。

重要名词和术语解析

Lipid: Lipids are substances that can be extracted from plant or animal matter by organic

脂质：脂质是可以通过有机溶剂（如氯仿、二乙醚或甲苯）

solvents, such as chloroform, diethyl ether, or toluene.

Triglyceride: The most common lipids are fats and oils composed of triglycerides formed from the alcohol glycerol, $CH_2(OH)CH(OH)CH_2OH$, and a long-chain fatty acid such as stearic acid, $CH_3(CH_2)_{16}COOH$.

Phosphoglyceride: An important class of lipids consists of **phosphoglycerides** (glycerophosphatides), which may be regarded as triglyderides in which one of the acids bonded to glycerol is orthophosphoric acid.

5.1.10 Enzymes
酶

从动植物中提取的物质。

甘油三酸酯：最常见的脂质有甘油三醇$CH_2(OH)CH(OH)CH_2OH$和长链脂肪酸，如硬脂酸$CH_3(CH_2)_{16}COOH$，组成的甘油三酸酯。

磷酸甘油酯：一类重要的脂质由**磷酸甘油酯**组成，可以认为是甘油三酸酯，其中与甘油键合的一种酸是正磷酸。

知识框架

Aspects of Enzymes

(1) Enzymatic reactions are reversible.

(2) Enzymes are named for where they are and what they do.

Gastric proteinase is a stomach enzyme that hydrolyzes proteins.

(3) Some enzymes require coenzymes in order to function.

Some vitamins are coenzymes.

(4) There are several important classes of enzymes:

1) Hydrolyzing enzymes cleave molecules and add H_2O.

2) Some enzymes break C—C bonds in chains of C atoms.

3) *Oxidases* and *reductases* enable oxidation and reduction.

4) *Isomerases* form isomers (such as

酶的性质

(1) 酶反应是可逆的。

(2) 酶因其所在位置和作用而命名。

胃蛋白酶是在胃里的蛋白酶，可水解蛋白质。

(3) 有些酶需要辅酶才能起作用。

一些维生素是辅酶。

(4) 几种重要的酶类：

1) 水解酶裂解分子并添加H_2O。

2) 一些酶破坏C原子链中的C—C键。

3) 氧化酶和还原酶可以氧化和还原。

4) 异构酶形成异构体

fructose to glucose).

5) Lyase enzymes remove chemical groups without hydrolysis.

6) *Ligase* enzymes link molecules, typically with C—C bonds.

(5) Enzyme action affected by many factors:

1) Temperature.

2) pH.

3) Toxic substances.

（如从果糖到葡萄糖）。

5）裂解酶去除化学基团而不水解。

6）连接酶通常通过 C—C 键连接分子。

（5）酶的作用受许多因素影响：

1）温度；

2）pH 值；

3）有毒物质。

重要名词和术语解析

Enzyme: Catalysts are substances that speed up a chemical reaction without themselves being consumed in the reaction. The most sophisticated catalysts of all are those found in living systems. They bring about reactions that could not be performed at all, or only with great difficulty, outside a living organism. These catalysts are called **enzymes**.

Substrate: Enzymes are proteinaceous substances with highly specific structures that interact with particular substances or classes of substances called **substrates**.

Enzyme-substrate complex: An enzyme "recognizes" a particular substrate by its molecular structure and binds to it to produce an **enzyme-substrate complex**.

Coenzyme: Some enzymes cannot function by themselves. In order to work, they must first be attached to **coenzymes**.

Hydrolyzing enzyme: Hydrolyzing enzymes

酶：催化剂是可加速化学反应而不被消耗的物质。最复杂的催化剂是生命系统中发现的那些催化剂。它们在生物体之外根本无法发生反应，或者很难发生反应。这些催化剂称为**酶**。

底物：酶是具有高度特定结构的蛋白质物质，可与特定物质或物质类型相互作用，这类物质称为**底物**。

酶-底物复合物：酶通过其分子结构"识别"特定底物，并与其结合以产生**酶-底物复合物**。

辅酶：某些酶本身不能起作用。为了起作用，它们必须首先与**辅酶**结合。

水解酶：水解酶通过加成水来

bring about the breakdown of high-molecular-weight biological compounds by the addition of water.

Oxidase: The enzymes that bring about oxidation in the presence of free O_2 are called **oxidases**. In general, biological oxidation-reduction reactions are catalyzed by **oxidoreductase enzymes**.

Isomerase: Isomerases form isomers of particular compounds.

Transferase enzyme: Transferase enzymes move chemical groups from one molecule to another.

Lyase enzyme: Lyase enzymes remove chemical groups without hydrolysis and participate in the formation of C=C bonds or addition of species to such bonds.

Ligase enzyme: Ligase enzymes work in conjunction with ATP (adenosine triphosphate, a high-energy molecule that plays a crucial role in energy-yielding, glucose-oxidizing metabolic processes) to link molecules together with the formation of bonds such as carbon-carbon or carbon-sulfur bonds.

5.1.11 Nucleic Acids
核酸

分解高分子量生物化合物。

氧化酶：在游离 O_2 存在下引起氧化的酶称为**氧化酶**。通常生物氧化还原反应由**氧化还原酶**催化。

异构酶：异构酶形成特定化合物的异构体。

转移酶：转移酶将化学基团从一个分子转移到另一分子。

裂解酶：裂解酶不经水解即可除去化学基团，并参与 C=C 键的形成或向该键添加物质。

连接酶：连接酶将 ATP（三磷酸腺苷，一种高能分子，在产生能量的葡萄糖氧化代谢过程中起着至关重要的作用）与分子连接在一起，并形成如碳-碳或碳-硫键之类的键。

知识框架

Nucleic acids (DNA and RNA) are enormous macromolecules composed of units of repeating units of nucleotides, each composed of a sugar, a nitrogenous base, and phosphate:

核酸（DNA 和 RNA）是巨大的大分子，由核苷酸的重复单元组成，每个重复单元由糖、含氮碱基和磷酸盐组成：

1) DNA acts like a coded message to pass on genetic information DNA.

2) RNA acts to synthesize proteins RNA.

The insight provided by this structure first proposed in 1953 enabled a revolution in biology leading to such things as recombinant DNA, fingerprinting of DNA, and advances still ongoing at a rapid pace.

Modified DNA

DNA may be modified by adding, deleting, or substituting nucleotides:

(1) Result may be a mutation transmittable to offspring.

(2) Toxicants may cause mutations.

(3) Radiation may cause mutations.

(4) Damage to DNA can result in cancer or birth defects.

1) 传递遗传信息的编码信息；

2) 合成蛋白质。

1953年首次提出这种结构所提供的见解促进了生物学革命，从而引发了重组DNA，DNA指纹图谱以及仍在快速发展中的进步。

修饰的DNA

可通过添加、删除或取代核苷酸来修饰DNA：

（1）结果可能是可传播给后代的突变；

（2）有毒物质可能导致突变；

（3）辐射可能导致突变；

（4）DNA受损可能导致癌症或先天缺陷。

重要名词和术语解析

Nucleic acid: The **nucleic acids** are macromolecules in living cells that serve to direct the function of the cells and cell organism reproduction.

Deoxyribonucleic acid (DNA): DNA contains the directions of cell function.

Ribonucleic acid (RNA): RNA carries out the directions given by DNA, primarily through directing the synthesis of various proteins in an organism.

Nucleoside: Each of the polymeric macromolecules is composed of many **nucleoside** groups linked in chains.

核酸：核酸是活细胞中的大分子，用于指导细胞的功能和细胞有机体的繁殖。

脱氧核糖核酸（DNA）：DNA包含细胞功能的指导。

核糖核酸（RNA）：RNA执行DNA给出的指导，主要是通过指导生物体内各种蛋白质的合成。

核苷：高分子大分子中的每一个均由许多链状连接的**核苷**基团组成。

Gene: The directions for the synthesis of a single protein are contained in a segment of DNA called a **gene**.

Transcription and **translation**: Portions of the DNA double helix unravel, and one of the strand of DNA produces a strand of RNA. This messenger RNA then acts as a template to assemble amino acids in the correct sequence for a particular protein in a process called **translation**.

Mutation: DNA molecules may be modified by the unintentional addition or deletion of nucleotides or by substituting one nucleotide for another. The result is a **mutation** that is transmittable to offspring.

基因：单一蛋白质的合成包含在称为**基因**的 DNA 片段中。

转录和翻译：DNA 的双螺旋部分解开，其中一条 DNA 链产生一条 RNA 链。然后，这种信使 RNA 充当模板，以特定的蛋白质的正确序列组装氨基酸，该过程称为**翻译**。

突变：DNA 分子可以通过无意添加或删除核苷酸，或用一个核苷酸替换为另一个而改变，造成可传播给后代的**突变**。

5.1.12　Recombinant DNA and Genetic Engineering　重组 DNA 与基因工程

知识框架

Recombinant DNA involves transfer of DNA material between species.

(1) May enable synthesis of protein not normally made by the organism.

1) Example: Human insulin biosynthesized by genetically engineered bacteria.

2) Example: Plants that produce biobased insecticide originally produced by *Bacillus thuringiensis* bacteria from which genes have been put into plants.

(2) Genetic engineering has numerous possibilities for green chemical synthesis.

Example: Lactic acid made by genetically

重组 DNA 涉及物种间 DNA 物质的转移。

(1) 可以合成生物体通常不合成的蛋白质。

1) 例：基因改造过的细菌生物合成的人胰岛素。

2) 例：将可产生生物基杀虫剂的苏云金芽孢杆菌的基因植入植物中，植物可产生生物基杀虫剂。

(2) 基因工程为绿色化学合成提供了多种可能性。

例：通过基因工程玉米制

engineered corn from which polylactic acid polymer may be synthesized.

得的乳酸，可以从中合成聚乳酸聚合物。

重要名词和术语解析

<u>Genetic engineering</u>：It has become possible to transfer the information between organisms by means of **recombinant DNA technology**, which has resulted in a new industry based on **genetic engineering**.

<u>基因工程</u>：通过**重组 DNA 技术**使生物体之间传递信息成为可能，这导致新产业**基因工程**的诞生。

<u>Cloning vehicles</u>：Molecules, called **cloning vehicles**, are capable of penetrating the host cell and becoming incorporated into its genetic material.

<u>克隆载体</u>：称为**克隆载体**的分子能够穿透宿主细胞并整合到其遗传物质中。

5.1.13 Metabolic Processes
代谢过程

知识框架

（1）Metabolism refers to biochemical processes that involve the alteration of biomolecules.

（1）代谢是指涉及生物分子改变的生化过程。

1）Anabolism (synthesis of larger molecules).

1）合成代谢（合成更大的分子）。

2）Catabolism (breakdown of molecules).

2）分解代谢（分子分解）。

（2）Energy-Yielding and Conversion Metabolic Processes.

（2）能量转化和代谢过程。

1）Respiration：Breakdown of molecules for energy.

1）呼吸：分解分子产生能量。

①Anoxic in the absence of molecular O_2.

①缺氧，没有分子氧的情况。

②Oxic, which uses molecular O_2：

②好氧，使用分子氧：

$$C_6H_{12}O_6(glucose，葡萄糖) + 6O_2 \longrightarrow 6CO_2 + energy，能量$$

2) Fermentation, such as ethanol from sugar:

$$C_6H_{12}O_6 \longrightarrow 2CO_2 + 2C_2H_5OH$$

3) Photosynthesis in which solar energy is converted to chemical energy in carbohydrate:

$$6CO_2 + 6H_2O + h\nu \longrightarrow C_6H_{12}O_6 + 6O_2$$

2) 发酵，例如从糖产生乙醇：

$$C_6H_{12}O_6 \longrightarrow 2CO_2 + 2C_2H_5OH$$

3) 光合作用，其中太阳能转化为碳水化合物中的化学能：

$$6CO_2 + 6H_2O + h\nu \longrightarrow C_6H_{12}O_6 + 6O_2$$

重要名词和术语解析

Metabolism: Biochemical processes that involve the alteration of biomolecules fall under the category of **metabolism**. Metabolic processes may be divided into the two major categories of **anabolism** (synthesis) and **catabolism** (degradation of substances).

Respiration: Organic compounds undergo catabolism that requires molecular oxygen (**aerobic respiration**) or that occurs in the absence of molecular oxygen (**anaerobic repiration**). Aerobic respiration uses the **Krebs cycle** to obtain energy.

$$C_6H_{12}O_6 \text{（glucose，葡萄糖）} + 6O_2 \longrightarrow 6CO_2 + \text{energy，能量}$$

Fermentation: Fermentation differs from respiration in not having an electron transport chain.

Photosynthesis: Light energy captured by plant and algal chloroplasts is used to synthesize sugars from carbon dioxide and water.

Autotrophic: Plant cells, which use sunlight as a source of energy and CO_2 as a source of carbon, are said to be autotrophic.

代谢：涉及生物分子改变的生化过程属于**代谢**。代谢过程可分为**合成代谢**（合成）和**分解代谢**（物质降解）两大类。

呼吸：有机化合物分解代谢需要分子氧（**有氧呼吸**）或在没有分子氧的情况下发生分解（**厌氧呼吸**）。有氧呼吸利用**克雷布斯循环**获得能量。

发酵：发酵与呼吸的区别在于没有电子传输链。

光合作用：植物和藻类叶绿体捕获的光能用来从二氧化碳和水中合成糖。

自养：将太阳光用作能量来源并将 CO_2 用作碳源的植物细胞是自养的。

Heterotrophic: Animal cells must depend upon organic material manufactured by plants for their food. These are called heterotrophic cells.

异养：动物细胞必须依靠植物生产的有机物质作为食物。这些被称为异养细胞。

5.1.14 Metabolism of Xenobiotic Substances 异种物质的代谢

知识框架

（1）Xenobiotic compounds are those normally foreign to living organisms.

（2）Metabolic processes can detoxify some toxic substances.

（3）Metabolic processes can also convert non-toxic protoxicants to toxic species.

（4）Intermediate xenobiotic metabolism can form transient species that may have particularly significant toxicological effects.

（1）Xenobiotic species may be acted upon by several different tissues and organs.

1）Often at entry site such as skin or lungs.

2）Liver is a particularly important organ for xenobiotic metabolism and may be damaged in the process.

（2）Phase I and Phase II reactions.

1）Phase I reaction normally introduces a functional group.

2）Phase II reaction attaches an endogenous conjugating agent to produce a conjugation product that is excreted.

（1）异源生物化合物是指通常对生物体来说为异物的化合物。

（2）代谢过程可以使某些有毒物质去毒。

（3）代谢过程还可以将无毒的前毒素转化为有毒物质。

（4）异源生物中间代谢可形成具有特别重要毒理作用的瞬时物种。

（1）异种物质可能被几种不同组织和器官作用。

1）通常在皮肤或肺等部位进入。

2）肝脏是异源生物代谢特别重要的器官，在此过程中可能会受损。

（2）第一阶段和第二阶段反应。

1）第一阶段反应通常会引入一个官能团。

2）第二阶段反应会附着内源性结合剂，从而产生结合产物，并被排泄。

重要名词和术语解析

Protoxicant: When toxicants or their metabolic precursors (**protoxicants**) enter a living organism they may undergo several processes, including those that may make them more toxic or that detoxify them.

Xenobiotic compound: Xenobiotic compounds are those that are normally foreign to living organisms; on chemical aspects; and on processes that lead to products that can be eliminated from the organism.

Intermediary xenobiotic metabolism: Intermediary xenobiotic metabolism results in the formation of somewhat transient species that are different from both those ingested and the ultimate product that is excreted.

Endogenous substrate: Xenobiotic compounds in general are acted upon by enzymes that function on a material that is in the body naturally—an endogenous substrate.

Biotransformation: Biotransformation refers to changes in xenobiotic compounds as a result of enzyme action.

Nonpolar lipophilic compound: nonpolar lipophilic compounds are relatively less soluble in aqueous biological fluids and more attracted to lipid species.

Phase Ⅰ reaction: A Phase Ⅰ reaction introduces reactive, polar functional groups into lipophilic ("fat-seeking") toxicant molecules.

前毒物：当毒物或其代谢前体（前毒物）进入生物时，它们可能会经历多个过程，包括可能使其毒性更大或使之解毒的过程。

异源化合物：异源化合物是通常对活生物体而言是外来的化合物；在化学方面；导致可以从生物体中清除产物的过程。

中间异源生物代谢：中间异源生物代谢导致形成某种瞬时物种，该物种既不同于摄入的物种，也不同于最终排出的产品。

内源性底物：异源化合物通常受天然作用于体内物质（内源性底物）上的酶作用。

生物转化：生物转化是指由于酶作用导致异源化合物的变化。

非极性亲脂性化合物：非极性亲脂性化合物在水性生物流体中的溶解度相对较低，并且更容易吸引脂质。

第一阶段反应：第一阶段反应将反应性极性官能团引入亲脂性（"寻找脂肪"）毒物分子中。

Phase Ⅱ reaction: The binding of such a substrate is a **Phase Ⅱ reaction**, and it produces a **conjugation product** that is amenable to excretion from the body.

第二阶段反应：这种底物的结合是**第二阶段反应**，它会产生**结合产物**，该结合产物易于从体内排泄。

5.2　Toxicological Chemistry
毒理化学

5.2.1　Introduction to Toxicology and Toxicological Chemistry
毒理学与毒理化学概论

知识框架

（1）Toxicology is the science of poisons.

（2）Toxicity depends upon many factors：

1）Degree of exposure.

2）Site of exposure.

3）Form of toxicant.

4）Dose.

5）Toxicant concentration.

6）Duration of exposure.

7）Rate of exposure.

（3）Four major types of exposure.

1）Acute：High exposure over a short time period.

① Local affecting the site of exposure.

② Systemic spread throughout organism.

2）Chronic：Generally low prolonged or repeated exposure.

① Local.

② Systemic.

（1）毒理学是毒药的科学。

（2）毒性取决于许多因素：

1）暴露程度；

2）暴露部位；

3）有毒物质的形式；

4）剂量；

5）毒物浓度；

6）暴露时间；

7）暴露率。

（3）四种主要暴露类型。

1）急性：短时间内高暴露。

①局部：影响暴露点。

②全身：在整个生物体中的扩散。

2）慢性：通常较低的长期或反复接触。

①局部。

②全身。

(4) Barriers to exposure.

1) Skin is an effective barrier in most locations of the body.

2) Skin in genital area more vulnerable.

Synergism, Potentiation, Antagonism of Toxicants

Chemical interactions between toxicants affect toxicities.

(1) Synergistic effects when the overall effect of two substances is greater than the sum of their parts.

(2) Potentiation when a substance inactive by itself enhances the action of an active substance.

(3) Antagonism when an active substance decreases the effect of another active one.

（4）暴露屏障。

1）皮肤是人体大部分部位的有效屏障。

2）生殖器部位的皮肤更脆弱。

毒物的协同、增强、拮抗作用

毒物之间的化学相互作用影响毒性。

（1）协同作用：两种物质的总作用大于其总和。

（2）增强作用：一种自身不活跃的物质增强活性物质的活性。

（3）拮抗作用：一种活性物质降低另一种活性物质的作用。

重要名词和术语解析

Toxicant: A poison, or toxicant, is a substance that is harmful to living organisms because of its detrimental effects on tissues, organs, or biological processes.

Toxicology: Toxicology is the science of poisons.

Matrix: A substance with which the toxicant is associated (the solvent in which it is dissolved or the solid medium in which it is dispersed) is called the matrix.

Acute local: Acute local exposure occurs at a

毒物：毒物是一种有害生物，因为它对组织、器官或生物过程具有有害作用。

毒理学：毒理学是毒药的科学。

基质：与毒物相关的物质（溶解有毒物的溶剂或分散有毒物的固体介质）称为基质。

急性局部暴露：急性局部暴露

specific location over a time period of a few seconds to a few hours and may affect the exposure site, particularly the skin, eyes, or mucous membranes.

Chronic local exposure: the time span may be as long as several years.

Acute systemic: Acute systemic exposure is a brief exposure or exposure to a single dose and occurs with toxicants that can enter the body, such as by inhalation or ingestion, and affect organs, such as the liver, that are remote from the entry site.

Chronic systemic: The exposure occurs over a prolonged time period.

Stratum corneum: The major barrier to dermal absorption of toxicants is the stratum corneum, or horny layer.

Biomonitor: Organisms can serve as indicators of various kinds of pollutants. In this application, organisms are known as biomonitors.

Additive: When both substances have the same physiologic function, their effects may be simply additive.

Synergistic: The total effect is greater than the sum of the effects of each separately.

Potentiation: Potentiation occurs when an inactive substance enhances the action of an active one.

Antagonism: Antagonism occurs when an active substance decreases the effect of another active one.

发生在几秒钟到几小时的特定位置，可能会影响暴露部位，尤其是皮肤、眼睛或黏膜。

慢性局部暴露：时间跨度可能长达数年。

急性全身性暴露：急性全身性暴露是短暂的暴露或单次暴露，并伴随有可能通过吸入或食入进入体内并影响远离进入部位的器官（如肝脏）的毒物。

慢性全身性暴露：暴露时间较长。

角质层：皮肤吸收有毒物质的主要障碍是角质层。

生物监测器：生物可以作为各种污染物的指标。这种生物被称为生物监测器。

加成作用：当两种物质具有相同的生理功能时，它们的共同作用可能只是加成。

协同作用：总效果大于每个单独效果的总和。

增强作用：当非活性物质增强活性物质的作用时，就会发生增强作用。

拮抗作用：当一种活性物质降低另一种活性物质的作用时，发生拮抗作用。

5.2.2 Dose-Response Relationships
剂量反应关系

知识框架

(1) Dose is the amount of substance to which an organism is exposed.

Usually amount per unit body mass.

(2) Response is what happens to an organism that is exposed.

1) Death is a common response in test organisms.

2) LD_{50} is the lethal dose for 50% of particular organisms.

(1) 剂量是生物体所接触的物质的量。

通常每单位体重的量。

(2) 反应是暴露的生物体的反应。

1) 死亡是测试生物中的常见反应。

2) LD_{50} 是 50% 特定生物的致死剂量。

重要名词和术语解析

Dose: Dose is the amount, usually per unit body mass, of a toxicant to which an organism is exposed.

Response: Response is the effect upon an organism resulting from exposure to a toxicant.

剂量：剂量是生物体所接触的有毒物质的量，通常是每单位体重。

反应：反应是由于接触毒物而对生物造成的影响。

5.2.3 Relative Toxicities
相对毒性

知识框架

Nonlethal Effects

Nonlethal effects and margin of safety are especially important for pharmaceuticals.

非致命效应

非致命作用和安全裕度对于药物尤为重要。

重要名词和术语解析

Potent: When there is a substantial difference between LD_{50} values of two different substances, the one with the lower value is said to be the more **potent**.

效力：当两种不同物质的 LD_{50} 值之间存在显著差异时，值较低的一种被认为是更**有效的**。

5.2.4 Reversibility and Sensitivity
可逆性和敏感性

知识框架

(1) Reversible effects leave no lasting effects.

(2) Irreversible effects have permanent effects.

(3) Hypersensitivity refers to individuals that respond to very low doses.

May be induced resulting in high sensitivity immune response.

(4) Hyposensitivity refers to subjects that can tolerate greater than normal doses.

For some substances tolerance may be built up.

(1) 可逆作用无持久影响。

(2) 不可逆作用具有永久的影响。

(3) 超敏反应是指对极低剂量有反应的个体。

可能被诱导导致高灵敏度的免疫反应。

(4) 低敏性是指可以耐受高于正常剂量的受试者。

对于某些物质,可能会建立耐受性。

重要名词和术语解析

Reversible: If there is no lasting effect from the exposure, it is said to be **reversible**. However, if the effect is permanent, it is termed **irreversible**.

Hypersensitivity and **hyposensitivity**: some subjects are very sensitive to a particular poison (for example, those killed at a dose corresponding to LD_5), whereas others are very resistant to the same substance (for example, those surviving a dose corresponding to LD_{95}). These two kinds of responses illustrate **hypersensitivity** and **hyposensitivity**, respectively; subjects in the mid-range of the dose-response curve are termed **normals**.

可逆的:如果暴露没有持久的影响,则认为是**可逆的**。但是如果影响是永久性的,则称为**不可逆**。

超敏反应和低敏反应:有些受试者对特定的毒物非常敏感(如以相当于LD_5的剂量杀死的受试者),而另一些受试者对同一物质具有非常的抵抗力(如那些存活于LD_{95}的剂量的受试者)。这两种反应分别说明了**超敏反应**和**低敏反应**。剂量反应曲线中段的受试者称为**正常受试者**。

5.2.5 Xenobiotic and Endogenous Substances
异源和内源性物质

知识框架

(1) Xenobiotic substances are foreign to living systems.

(2) Endogenous substances are normally present in biological systems.

Often required at lower levels, toxic at higher levels.

(3) Calcium illustrates the behavior shown above with toxic and even lethal effects both for deficiency and in excess.

(1) 异源物质对生命系统是异物。

(2) 内源性物质通常存在于生物系统中。

通常少量需要，在较高水平时有毒。

(3) 缺乏和过量都有毒甚至致命作用。

重要名词和术语解析

Xenobiotic：Xenobiotic substances are those that are foreign to a living system, whereas those that occur naturally in a biologic system are termed **endogenous**.

异源生物：异源生物物质是指生命系统中不存在的物质，而生物系统中天然存在的物质则称为**内源性物质**。

5.2.6 Toxicological Chemistry
毒理化学

知识框架

Toxicological chemistry deals with the chemical nature and reactions of toxic substances including:

(1) Origins.

(2) Uses.

(3) Chemical aspects of exposure, fates, disposal.

Most toxic substances and their precursors (protoxicants) are metabolized, commonly by

毒理化学涉及有毒物质的化学性质和反应，包括：

(1) 来源；

(2) 用途；

(3) 化学暴露、归宿、处置。

大多数有毒物质及其前体（前毒素）被代谢，通常通过

Phase I and Phase II Reactions.

(1) Designed to detoxify.

(2) Product is usually more readily eliminated.

(3) May produce a toxic or more toxic form.

(1) Most Phase I reactions are catalyzed by microsomal mixed-function oxidase.

(2) Reactions.

1) Catalyzed by the cytochrome P-450 enzyme system.

2) Associated with cell endoplasmic reticulum.

3) Occurring most abundantly in the liver of vertebrates.

第一阶段和第二阶段反应进行代谢。

(1) 专为排毒而设计。

(2) 产物通常更容易去除。

(3) 可能产生有毒或更有毒的形式。

(1) 大多数第一阶段反应均由微粒体混合功能氧化酶催化。

(2) 反应。

1) 由细胞色素P-450酶系统催化。

2) 与细胞内质网相关。

3) 最常发生在脊椎动物的肝脏中。

重要名词和术语解析

Toxicological chemistry: Toxicological chemistry is the science that deals with the chemical nature and reactions of toxic substances, including their origins, uses, and chemical aspects of exposure, fates, and disposal.

Phase I reaction: Lipophilic xenobiotic species in the body tend to undergo **Phase I reactions** that make them more water-soluble and reactive by the attachment of polar functional groups, such as —OH. Most Phase I processes are "microsomal mixed-function oxidase" reactions catalyzed by the cytochrome P-450 enzyme system associated with the **endoplasmic reticulum** of the cell and occurring most abundantly in the liver of vertebrates.

毒理化学：毒理化学是一门研究有毒物质的化学性质和反应的科学，包括其来源、用途以及暴露、归宿和处置的化学方面。

第一阶段反应：体内的亲脂性异种生物倾向于发生**第一阶段反应**，这些反应通过与极性官能团（如—OH）的连接使它们更具水溶性和反应性。大多数第一阶段过程是与细胞**内质网**相关的细胞色素P-450酶系统催化的"微粒体混合功能氧化酶"反应，在脊椎动物的肝脏中含量最高。

Phase Ⅱ reaction: A Phase Ⅱ reaction occurs when an endogenous species is attached by enzyme action to a polar functional group which often, though not always, is the result of a Phase Ⅰ reaction on a xenobiotic species.

Conjugation reaction: Phase Ⅱ reactions are called **conjugation reactions** in which enzymes attach **conjugating agents** to xenobiotics, their Phase Ⅰ reaction products, and nonxenobiotic compounds. The **conjugation product** of such a reaction is usually less toxic than the original xenobiotic compound, less lipid-soluble, more water-soluble, and more readily eliminated from the body.

第二阶段反应：当内源性物质通过酶作用与极性官能团连接时，就会发生第二阶段反应，这通常（尽管并非总是如此）是异源生物物种发生第一阶段反应的结果。

缀合反应：第二阶段反应称为**缀合反应**，其中酶将**缀合剂**连接到异种生物，它们的第一阶段反应产物和非异种生物化合物上。这种反应的**结合产物**通常比原始的异种生物化合物毒性低，脂溶性较小，水溶性更大，更容易从体内清除。

5.2.7 Kinetic Phase and Dynamic Phase 动力学阶段和动态阶段

知识框架

(1) In the kinetic phase a toxicant or precursor to a toxicant (protoxicant) may undergo absorption, metabolism, temporary storage, distribution and excretion.

(2) In the dynamic phase a toxicant or toxic metabolite interacts with cells, tissues, or organs to cause a toxic response.

Example of Dynamic Phase Process

(1) Primary reaction (carbon monoxide with blood hemoglobin):

$$O_2(Hb,血红蛋白)+CO \Longrightarrow CO(Hb,血红蛋白)+O_2$$

(2) Biochemical response:
Deprivation of O_2 to tissue.

(1) 在动力学阶段，有毒物质或有毒物质的前体（前毒物）可能会发生吸收、代谢、暂时存储、分配和排泄。

(2) 在动态阶段，有毒物或有毒代谢产物与细胞、组织或器官相互作用，引起中毒反应。

动态阶段过程的例子

(1) 原发反应（一氧化碳与血红蛋白）：

(2) 生化反应：
剥夺组织中的氧气。

(3) Observable effect:
Lowered consciousness>coma>death.

(3) 可见效果：
意识下降>昏迷>死亡。

重要名词和术语解析

Kinetic phase: In the **kinetic phase**, a toxicant or the metabolic precursor of a toxic substance (**protoxicant**) may undergo absorption, metabolism, temporary storage, distribution, and excretion.

Active parent compound: A toxicant that is absorbed may be passed through the kinetic phase unchanged as an **active parent compound**, metabolized to a **detoxified metabolite** that is excreted, or converted to a toxic **active metabolite**.

Dynamic phase: In the **dynamic phase** a toxicant or toxic metabolite interacts with cells, tissues, or organs in the body to cause some toxic response.

动力学阶段：在**动力学阶段**，有毒物质或有毒物质的代谢前体（**前毒物**）可能会发生吸收、代谢、暂时存储、分配和排泄。

活性母体化合物：被吸收的毒物可以不经过**活性母体化合物**而通过动力学阶段，代谢成排泄的**解毒代谢物**，或转化为**有毒活性代谢物**。

动态阶段：在**动态阶段**，有毒物或有毒代谢产物与体内的细胞、组织或器官相互作用，引起某些毒性反应。

5.2.8 Teratogenesis, Mutagenesis, Carcinogenesis, and Effects on the Immune and Reproductive Systems
致畸、致突变、致癌作用及其对免疫和生殖系统的影响

知识框架

(1) These are some of the most common effects of exposure to toxic substances.

(2) Carcinogenesis often receives the most attention.

(1) 这些是接触有毒物质最常见的影响。

(2) 致癌作用最受关注。

Teratogenesis

(1) Teratogens are chemical species that cause birth defects.

致畸作用

(1) 致畸物是导致出生缺陷的化学物质。

1) Usually from damage to embryonic or fetal cells.

2) Mutations in egg or sperm cells.

(2) Biochemical mechanisms of teratogenesis from exposure to toxicants include:

1) Enzyme inhibition.

2) Deprivation of essential substrates.

3) Interference with energy supply.

4) Alteration of permeability of placental membrane.

(3) Fetuses exposed to toxic substances *in utero* are the most vulnerable to the effects of xenobiotics.

Mutagenesis

(1) Mutagens alter DNA to produce inheritable traits.

1) Mutagens are also likely to cause cancer.

2) Mutagens may cause birth defects.

(2) Most mutations from toxicants result from alteration of DNA such as by alkylation.

Carcinogenesis

(1) Cancer is the uncontrolled replication and growth of the body's own cells.

(2) Carcinogenic agents are primarily:

1) Chemical agents, such as nitrosamines and polycyclic aromatic hydrocarbons;

2) Biological agents, such as hepadnaviruses

致突变作用

(1) 突变基因改变DNA以产生可遗传的特征。

1) 突变也可能导致癌症;

2) 突变可能导致先天缺陷。

(2) 毒药的大多数突变是由DNA的改变引起的,例如通过烷基化。

致癌作用

(1) 癌症是人体自身细胞不受控制的复制和生长。

(2) 致癌剂主要是:

1) 化学试剂,比如亚硝胺和多环芳烃;

2) 生物制剂,比如肝炎

通常是因为胚胎或胎儿细胞受损。

2) 卵子或精子细胞发生突变。

(2) 接触毒物致畸的生化机制包括:

1) 抑制酶;

2) 剥夺必需底物;

3) 干扰能源供应;

4) 胎盘膜通透性的改变。

(3) 胎儿在子宫内接触有毒物质最容易受到异源生物的影响。

or retroviruses.

3) Ionizing radiation, such as X-rays.

4) Genetic factors, such as selective breeding.

(1) Toxicological chemistry is primarily concerned with chemical carcinogenesis.

(2) History of chemical carcinogenesis:

1) 1775 Sir Percival Pott noted cancer of the scrotum in chimney sweeps.

2) Around 1900 a German surgeon, Ludwig Rehn, reported elevated incidences of bladder cancer in dye workers exposed to chemicals extracted from coal tar; 2-naphthylamine.

3) Observations of cancer from tobacco juice (1915).

4) Oral exposure to radium from painting luminescent watch dials.

5) Tobacco smoke (1939).

6) Asbestos (1960).

7) Vinyl chloride (around 1970).

Initiation stage of carcinogenesis

(1) DNA-reactive species (commonly alkylating agents).

(2) Epigenetic carcinogen.

Promoters act after initiation.

Alkylating agents in carcinogenesis

Bruce Ames test to infer carcinogenicity

Reversion of histidine-requiring *Salmonella* bacteria to a form that can synthesize histidine.

Histidine-requiring *Salmonella* mixed with

癌变的起始阶段

（1）DNA 反应物种（通常为烷基化剂）。

（2）表观遗传致癌物致癌。

启动子发动后就作用。

在致癌作用中的烷基化剂

Bruce Ames 测试推断致癌性

将需要组氨酸的沙门氏菌还原为可以合成组氨酸的形式。

将需要组氨酸的沙门氏菌

liver homogenate inoculated onto histidine-free media:

(1) Liver homogenate has enzymes that can convert procarcinogens to carcinogens.

(2) Growth of *Salmonella* shows reversion back to wild strain indicative of mutagen (inferred carcinogen).

Immune System Response

(1) Immune system protects against:

1) Xenobiotic chemicals.

2) Infectious agents.

3) Neoplastic cells leading to cancer.

(2) Immunosuppression by toxicants can reduce effectiveness of the immune system.

(3) Hypersensitivity (allergy) is overreaction of immune system that can be caused by foreign agents such as:

1) Beryllium.

2) Chromium.

3) Nickel.

4) Formaldehyde.

5) Some kinds of pesticides, resins, and plasticizers.

Endocrine Disruption

(1) Endocrine gland activities regulate.

1) Metabolism.

2) Reproductive function.

(2) Aquatic organisms particularly susceptible to endocrine disrupting chemicals in water.

1) Reproductive dysfunction.

2) Abnormal serum steroid levels.

与肝匀浆混合接种到无组氨酸的培养基上:

(1) 肝匀浆具有可将致癌物转化为致癌物的酶;

(2) 沙门氏菌的生长表明已回复到指示诱变剂(指向致癌物)的野生菌株。

免疫系统反应

(1) 免疫系统可防止:

1) 异源化学物质;

2) 传染源;

3) 肿瘤细胞导致癌症。

(2) 有毒物质的免疫抑制可降低免疫系统的有效性。

(3) 超敏反应(过敏)是免疫系统的过度反应,由以下物质引起:

1) 铍;

2) 铬;

3) 镍;

4) 甲醛;

5) 某些农药、树脂和增塑剂。

内分泌干扰

(1) 内分泌活动调节:

1) 代谢;

2) 生殖功能。

(2) 水生生物特别易受内分泌干扰物破坏生殖功能障碍。

1) 血清类固醇水平异常;

2) 性别特征的改变;

3) Alterations in sex characteristics.

(3) Hormonally-active agents, especially those that act like estrogen, female sex hormone.

1) Estrogen.

2) Synthetic 17a-ethinylestradiol.

(4) Xenoestrogens are estrogenic substances from artificial sources.

Endocrine disruptors that may affect organisms in the environment. These substances include:

(1) Natural hormones.

(2) Synthetic hormones in oral contraceptives.

(3) Breakdown products of surfactants.

(4) Epoxy resin ingredients.

(5) Polymer plasticizers.

(6) Substances made by trees to resist disease.

3) 激素活性剂。

(3) 特别是那些作用像雌激素，雌激素的激素活性剂。

1) 雌激素；

2) 合成的17a-炔雌醇。

(4) 异雌激素是来自人工来源的雌激素物质。

可能影响环境中生物的内分泌干扰物包括：

(1) 天然激素；

(2) 口服避孕药中的合成激素；

(3) 表面活性剂的分解产物；

(4) 环氧树脂成分；

(5) 聚合物增塑剂；

(6) 树木制成的抗病物质。

重要名词和术语解析

Teratogen: Teratogens are chemical species that cause birth defects.

Mutagen: **Mutagens** alter DNA to produce inheritable traits.

Alkylation: The attachment of a small alkyl group, such as —CH_3 or —C_2H_5, to an N atom on one of the nitrogenous bases in DNA is one of the most common mechanisms leading to mutation.

Chemical carcinogenesis: The role of xenobiotic chemicals in causing cancer is called **chemical carcinogenesis**.

致畸物：致畸物是导致出生缺陷的化学物质。

诱变剂：诱变剂会改变DNA产生可遗传的性状。

烷基化：将一个小烷基（如 —CH_3 或 —C_2H_5）连接到DNA中一个含氮碱基上的N原子上，是最常见的导致突变的机制之一。

化学致癌作用：异源化学物质在致癌中的作用称为**化学致癌作用**。

Initiation: Initiation of carcinogenesis may occur by reaction of a **DNA-reactive species** with DNA, or by the action of an **epigenetic carcinogen** that does not react with DNA and is carcinogenic by some other mechanism. Most DNA-reactive species are **genotoxic carcinogens** because they are also mutagens.

Procarcinogen: Cancer-causing substances that require metabolic activation are called **procarcinogens**.

Ultimate carcinogen: The metabolic species actually responsible for carcinogenesis is termed an **ultimate carcinogen**.

Proximate carcinogen: Some species that are intermediate metabolites between precarcinogens and ultimate carcinogens are called **proximate carcinogens**.

Primary or **direct-acting carcinogen**: Carcinogens that do not require biochemical activation are categorized as **primary** or **direct-acting carcinogens**.

Promoter: Most substances classified as epigenetic carcinogens are **promoters** that act after initiation.

Arylating agent: **Arylating agents** act to attach aryl moieties, such as the phenyl group to DNA.

Bruce Ames: Mutagenicity used to infer carcinogenicity is the basis of the **Bruce Ames** test, in which observations are made of the reversion of mutant histidine-requiring *Salmonella* bacteria back to a form that can synthesize its own histidine.

引发：可能通过 DNA 反应性物种与 DNA 反应，或通过不与 DNA 发生反应并通过其他某种机制致癌的**表观遗传致癌物**的作用来**引发**癌变。大多数 DNA 反应性物种都是**遗传毒性致癌物**，因为它们也是诱变剂。

致癌物：需要代谢激活的致癌物质称为**致癌物**。

最终致癌物：真正引起癌变的代谢物种称为**最终致癌物**。

近致癌物：在前致癌物和最终致癌物之间的中间代谢物的某些物种称为**近致癌物**。

原发性或直接作用的致癌物：不需要生化活化的致癌物归类为**原发性**或**直接作用的致癌物**。

启动子：大多数被分类为表观遗传致癌物的物质都是启动后起作用的**启动子**。

芳基化剂：芳基化剂将芳基部分（如苯基）连接到 DNA。

Bruce Ames：用于推断致癌性的诱变性是**布鲁斯·埃姆斯**（Bruce Ames）测试的基础，在该测试中，观察到突变的需要组氨酸沙门氏菌回到可以合成其自身组氨酸的形式。

Immune system: The **immune system** acts as the body's natural defense system to protect it from xenobiotic chemicals; infectious agents, such as viruses or bacteria; and neoplastic cells, which give rise to cancerous tissue.

Immunosuppression: Toxicants can cause **immunosuppression**, which is the impairment of the body's natural defense mechanisms.

Allergy or **hyper-sensitivity**: Another major toxic response of the immune system is **allergy** or **hyper-sensitivity**.

免疫系统：**免疫系统**是人体的天然防御系统，可保护人体免受异源化学物质、传染性物质（如病毒或细菌）和肿瘤细胞（这些细胞会引起癌变）的侵害。

免疫抑制：有毒物质可引起**免疫抑制**，这是人体自然防御机制的损害。

过敏或**超敏反应**：免疫系统的另一个主要毒性反应是**过敏**或**超敏反应**。

5.2.9 Health Hazards
健康危害

知识框架

Assessment of Potential Exposure

(1) Measure chemicals and their products in exposed subject.

(2) Epidemiological studies.

Clusters of spontaneous abortions, birth defects, particular kinds of cancers.

(3) Estimation of health effects risks.

Uncertainties in extrapolating from short-term acute effects of high doses to long-term chronic effects of low doses.

Health Risk Assessment

For example, of hazardous waste sites.

Risk assessment includes:

(1) Identification of hazard.

(2) Dose-response assessment.

(3) Exposure assessment.

(4) Risk characterization.

潜在暴露评估

（1）测量暴露对象中的化学物质及其产物。

（2）流行病学研究。

自然流产、先天缺陷、某些种类的癌症的集群。

（3）健康影响风险估计。

从高剂量的短期急性影响推断到小剂量的长期慢性影响的不确定性。

健康风险评估

例如危险废物场所。

风险评估包括：

（1）危害识别；

（2）剂量反应评估；

（3）暴露评估；

（4）风险特征。

重要名词和术语解析

Epidemiological study: **Epidemiological studies** applied to toxic environmental pollutants, such as those from hazardous wastes, attempt to correlate observations of particular illnesses with probable exposure to such wastes.

Cluster: **Clusters** consisting of an abnormally large number of cases of a particular disease in a limited geographic area.

Health risk assessment: One of the major ways in which toxicology interfaces with the area of hazardous wastes is in **health risk assessment**, providing guidance for risk management, cleanup, or regulation needed at a hazardous waste site based upon knowledge about the site and the chemical and toxicological properties of wastes in it.

流行病学研究：流行病学研究适用于有毒的环境污染物，如来自危险废物的污染物，试图将对特定疾病的观察结果与可能暴露于此类废物的结果联系起来。

集群：**集群**是由在有限地理区域内异常疾病的大量病例组成。

健康风险评估：健康风险评估是毒理学与危险废物领域相联系的主要方式之一，它基于危险废物现场、废物的化学和毒理特性的知识，为危险废物场所的风险管理、清理或法规提供指导。

Exercises
习题

5-1 What is the toxicological importance of lipids? How are lipids related to hydrophobic pollutants and toxicants?

5-2 What is the function of a hydrolase enzyme?

5-3 Match the cell structure with its function, below.
  (1) Mitochondria.
  (2) Endoplasmic reticulum.

5-1 脂类的毒理学意义是什么，脂类与疏水性污染物和毒物的关系如何？

5-2 水解酶的功能是什么？

5-3 将细胞结构与功能相匹配。
  (1) 线粒体。
  (2) 内质网。

(3) Cell membrane.
(4) Cytoplasm.
(5) Cell nucleus.

(A) Toxicant metabolism
(B) Fills the cell
(C) Deoxyribonucleic acid
(D) Mediate energy conversion and utilization
(E) Encloses the cell and regulates the passage of materials into and out of the cell interior

5-4 The formula of simple sugars is $C_6H_{12}O_6$. The simple formula of higher carbohydrates is $C_6H_{10}O_5$. Of course, many of these units are required to make a molecule of starch or cellulose. If higher carbohydrates are formed by joining together molecules of simple sugars, why is there a difference in the ratios of C, H, and O atoms in the higher carbohydrates as compared to the simple sugars?

5-5 Why does wood contain so much cellulose ($C_{600}H_{1000}O_{500}$)?

5-6 What would be the chemical formula of a trisaccharide made by the bonding together of three simple sugar molecules?

5-7 The general formula of cellulose may be represented as $(C_6H_{10}O_5)_x$. If the molar mass of a molecule of cellulose is 400000, what is the estimated value of x?

5-8 During one month a factory for the

（3）细胞膜。
（4）细胞质。
（5）细胞核。

（A）毒物新陈代谢
（B）填充细胞
（C）脱氧核糖核酸
（D）介导能量转换和利用
（E）封闭细胞，调节材料进入和离开细胞内部

5-4 单糖的简单公式是 $C_6H_{12}O_6$，高等碳水化合物的简单公式是 $C_6H_{10}O_5$。当然，制造一分子淀粉或纤维素需要许多这样的单位。如果说高等碳水化合物是由单糖分子连接在一起形成的，那么为什么高等碳水化合物中C、H、O原子的比例与单糖相比有差别呢？

5-5 为什么木材中含有这么多的纤维素（$C_{600}H_{1000}O_{500}$）？

5-6 由三个简单的糖分子结合在一起制成的三糖的化学式是什么？

5-7 纤维素的通式可表示为$(C_6H_{10}O_5)_x$。若一分子纤维素的摩尔质量为400000，则x的估计值为多少？

5-8 一个月内，一家生产单糖

production of simple sugars, $C_6H_{10}O_6$, by the hydrolysis of cellulose processes one million pounds of cellulose. The percentage of cellulose that undergoes the hydrolysis reaction is 40%. How many pounds of water are consumed in the hydrolysis of cellulose each month?

5-9 How are conjugating agents and Phase II reactions involved with some toxicants?

5-10 What is the toxicological importance of proteins, particularly as related to protein structure?

5-11 What is the toxicological importance of lipids? How are lipids related to hydrophobic pollutants and toxicants?

5-12 What are Phase I reactions? What enzyme system carries them out? Where is this enzyme system located in the cell?

5-13 What is a dose-response curve?

5-14 What are the three major subdivisions of the dynamic phase of toxicity, and what happens in each?

5-15 Characterize the toxic effect of carbon monoxide in the body. Is its effect reversible or irreversible? Does it act on an enzyme system?

5-16 Of the following, choose the one that is not a biochemical effect of a toxic substance (　　). (explain)

A. impairment of enzyme function by binding to the enzyme

B. alteration of cell membrane or carriers in cell membranes

C. change in vital signs

D. interference with lipid metabolism

E. interference with respiration

5-17 Distinguish among teratogenesis, mutagenesis, carcinogenesis, and immune system effects. Are there ways in which they are related?

5-18 As far as environmental toxicants are concerned, compare the relative importance of acute and chronic toxic effects and discuss the difficulties and uncertainties involved in studying each.

5-19 What are some of the factors that complicate epidemiologic studies of toxicants?

5-20 Alkylating agents do not or are not (　　). (explain)

A. formed by metabolic activation

B. attach groups such as CH_3 to DNA

C. include some species that cause cancer

D. alter DNA

E. noted for being electron-pair donors or nucleophiles

5-21 Of the following, if any, the untrue statement regarding Phase I reactions is (　　). (explain)

A. they tend to introduce reactive, polar functional groups onto lipophilic ("fat-seeking") toxicant molecules

B. the product of a Phase I reaction is usually more water-soluble than the parent xenobiotic species

B. 改变细胞膜或细胞膜中的载体

C. 改变生命体征

D. 干扰脂质代谢

E. 干扰呼吸

5-17 区分致畸、致突变、致癌和免疫系统效应，它们之间是否有关联？

5-18 就环境毒物而言，比较急性和慢性毒物效应的相对重要性，并讨论研究每种毒物所涉及的困难和不确定性。

5-19 使毒物流行病学研究复杂化的因素有哪些？

5-20 烷化剂不是或不会（　　）。（请解释）

A. 由代谢活化形成

B. 将 CH_3 等基团附着在 DNA 上

C. 包括一些致癌的物种

D. 改变 DNA

E. 因是电子对供体或亲核剂而受到注意

5-21 下列关于第一阶段反应的说法中，不正确的是（　　）。（请解释）

A. 它们往往在亲脂性（"寻求脂"）毒物分子上引入反应性的极性官能团

B. 第一阶段反应的产物通常比母体异源生物更易溶于水

C. the product of a Phase Ⅰ reaction possesses a "chemical handle" to which a substrate material in the body may become attached so that the toxicant can be eliminated from the body
D. Phase Ⅰ reactions are generally conjugation reactions through which an endogenous conjugating agent is attached
E. Phase Ⅰ reactions are catalyzed by enzymes

C. 第一阶段反应的产物具有"化学手柄",体内的底物可附着在该手柄上,以便将毒物排出体外

D. 第一阶段反应一般是共轭反应。通过其连接内源性结合剂

E. 第一阶段反应由酶催化

扫描二维码查看习题答案

参 考 文 献

[1] Stanley Manahan. Environmental Chemistry [M]. Tenth Edition. Los Angeles: CRC Press, 2017.
[2] 戴树桂. 环境化学 [M] 2 版. 北京: 高等教育出版社, 2006.
[3] Gary W VanLoon, Stephen J Dully. Environmental Chemistry [M]. Fourth Edition. Oxford: Oxford University Press, 2018.